The Painted Stork

A.J. Urfi

# The Painted Stork

## Ecology and Conservation

Springer

A.J. Urfi
Department of Environmental Biology
University of Delhi
New Delhi 110007
India
ajurfi@rediffmail.com

ISBN 978-1-4419-8467-8 e-ISBN 978-1-4419-8468-5
DOI 10.1007/978-1-4419-8468-5
Springer New York Dordrecht Heidelberg London

Library of Congress Control Number: 2011931533



Printed on acid-free paper

Springer is part of Springer Science+Business Media (www.springer.com)

*For Zara*
*And*
*For Siddiq Ahmed Siddiqi*

# Preface

*It is cheating to copy the work of another man; you should copy the work of TWO other men – they call that research (Love 1980; original source unknown)*

Writing a book has everything to do with ones involvement with his subject. My association with the Painted Stork spans over 20 years and has been influenced by a host of personal and professional factors. While I hope to write about these, elsewhere, someday, the significant point is that for studying these delectable birds I did not have to venture very far most of the time. In my home town Delhi – India's capital city, a colony of Painted Storks has been in existence since 1960. Each year in September, these tall, elegant birds start gathering at the Delhi Zoo where they build their nests on trees planted on islands in the ponds of the zoo. During the period of their residence, they breed, raise their young, oversee preliminary rites of passage and as soon as summer approaches, they are all gone, widely dispersed in the surrounding countryside, where they will live singly or in loose scattered parties seeking food and shelter for the rest of the year. Year after year I have seen the same pattern repeating itself.

The Painted Stork is a large, eye-catching, interesting looking, colonially nesting wading bird, listed as near threatened by international conservation agencies. It is found across large parts of South Asia and Southeast Asia, with a stronghold in the Indian subcontinent, particularly India and Sri Lanka. The genus *Mycteria*, to which it belongs, has representatives in three continents – Asia, America, and Africa, each differing from the other in minor details. But while these facts are interesting in themselves they are not sufficient to explain why a book on this species should be written. In Chap. 1 (Introduction), I have devoted a full section to explaining why a study of the Painted Stork and similar birds can be meaningful as well as topical. Therefore, while it will be pointless to repeat some of those points here, it should suffice to say, that, for me writing this book was a natural culmination of a journey begun long ago; a journey which has mostly been great fun and still continues.

British author and historian, William Dalrymple, well known for his books on Indian history, including several on Delhi, once said in an interview to a local newspaper,[1]

[1] The Indian Express, 7 February 1999.

> Books are sheer hard work, journalism less difficult, producing documentaries the easiest things to do. But the money is in inverse proportion.

Having written a few books before this one, I am somewhat inclined to agree with this view. Also, in the career of a university teacher, while research papers in trade journals often come in handy while applying for promotions and grants, scholarly books (at least in the science departments), not necessarily so.

Those readers who are familiar with the game of cricket, so popular in countries which were once a part of the erstwhile British Empire, may be aware that the game has two versions. The popular, faster, and more recent version is known as limited over's cricket while the classical version, which is long drawn and bereft of the sparkle and dazzle of one day or T-20 matches, is known as "test cricket." Connoisseurs of the game hold that test cricket is the noblest and purest form of the game and, in some sense, the ultimate challenge for a player. If that be so then I would like to think that writing a scholarly book is also a bit like playing test cricket in that, one gets a chance to develop the subject in a manner which is not possible in journal publications of restricted length. Perhaps, writing a species monograph is also the ultimate test for an author as he settles down to play a long innings, drawing upon the knowledge and wisdom accumulated over the years.

All these years, while I have been on the trail of the Painted Stork, the focus of my studies has largely been the Delhi Zoo population. Besides the fact that it is so easy and convenient to study these birds at close quarters, a major advantage, when I began my studies, was that the basic work had already been done on this population in that, during 1966–1971, J.H. Desai, a former director of the Delhi Zoo, had published a detailed study on various aspects of this population. His work is an impressive data base and during the course of my own studies I have drawn heavily from it. Sometimes, when I feel that in my studies of the Painted Stork I have explored hitherto untouched aspects of this population and in doing so contributed towards a better understanding of the biology of this species, I am reminded of the famous line attributed to Sir Issac Newton,

> If I have seen further it is only by standing on the shoulders of giants

I daresay that, due to our collective efforts, the Delhi Zoo Painted Stork population is perhaps the most well-studied among any population of (wild) birds in India (Urfi 2010). The two bursts of research activity on this population, first by Desai and coworkers and the other by my group have more or less coincided with other events happening in the ornithological world. For instance, the period of 1960s was one in which basic studies or whole biology studies of different species of wildlife were in vogue. This period also coincided with the considerable worldwide interest in the ecology, systematics and behavior of storks, and other waterbirds, as is amply clear from the publications of several international scientists cited in this book. The 1980s was a period in which considerable interest in field studies was generated by the initiation of Asian Waterfowl Census (Urfi et al. 2005). The easiest and most meaningful thing to do during those days was to count waterfowl and report findings, which is what I also initially did with the Painted Storks.

Modern biologists examine patterns of variability and ask specific questions about how a given trait could have evolved. An important arsenal in their research methodology is the comparative approach. In this book I have taken cognizance of research on other species of *Mycteria*, particularly the Wood Stork and Milky Stork and therefore employed the comparative approach freely. However, I must also confess that a bias towards Indian studies has been largely unavoidable, given that India is stronghold of the Painted Stork, and it is from here that most of the research on this species has emanated.

This book could have been written as an encyclopedic presentation of facts about the Painted Stork but then a very fundamental purpose – enthusing other field ornithologists to undertake field studies on similar birds, take a stronger interest about wetland ecology and conservation issues, etc., might have been defeated. Hence, I have tried to build this book mostly around the studies that have been undertaken on the Painted Stork, particularly those from Delhi. In writing this book I have also tried to review various points of interest from across the range of the Painted Stork in an attempt to give the topic a pan Asian coverage.

*Structure of the book*: Chapter 1 (Introduction) attempts what all introductory chapters are supposed to do i.e., introduce the subject, discuss its significance and outline the basic biology of the species by touching upon its systematics, morphology, life history, etc.

The fact that the Painted Stork is a colonially nesting species should suffice to explain why a chapter has been devoted to a breeding pattern, which is at best limited to only a small number of bird species (Chap. 2: Coloniality). The first part of this chapter serves to introduce the reader to the basic biological reasoning that has gone on to address the question of evolution of coloniality. However, a comprehensive review of all the theories about avian coloniality have not been undertaken here for which the reader will have to go though other sources. The second half of the chapter is devoted to describing the salient features of some of the important colonies which are referred to from time to time in the book.

Chapter 3 (Nesting Ecology) is devoted to different aspects of the breeding ecology of the Painted Stork, particularly the timing of reproduction and the role of proximate and ultimate factors. Food availability is widely regarded as the ultimate factor governing nesting time in birds and in the context of Asia which has a monsoonal climate, the whole issue has to be examined and understood in light of these factors. A review of the various biotic and abiotic factors that cause nest losses, at egg and nestling stage, is also included.

Due to the numerous studies undertaken on the Delhi Zoo Painted Stork population, this system is now fairly well understood. For instance, it is known that certain colonies are occupied first; which are those colonies? What are the patterns of occupation of colonies? And so on. Chapter 4 (Sexual Size Dimorphism and Mating Patterns) focuses on SSD, mating patterns, and habitat selection in the Painted Stork. The details of a study on SSD, using novel field study techniques are described in this chapter.

Chapter 5 (Growth and Development) is devoted to post-hatching growth and development of Painted Stork. Based largely on the studies conducted at the Delhi

Zoo by J.H. Desai and coworkers, an important consideration behind writing this chapter was to provide an illustrated guide about the characteristic growth stages of nestlings so that it may be possible to age the nestlings properly just by observing them through binoculars or a telescope in the field.

In birds the most important structure for efficient harvesting of food is the trophic apparatus, the bill. Some classical studies on Wood Stork demonstrated tactolocation in the *Mycteria* and these are reviewed in Chap. 6 (Food and Foraging). The bulk of the chapter revolves around our own studies on the foraging behavior of the Painted Stork in the Delhi region, covering aspects related to diet, length of bouts, prey sizes, seasonal changes in foraging patterns, etc. Also included in this chapter are the interactions of Painted Stork (and other heronry birds) with the environment, particularly enrichment of waters through dropping of guano and harvesting of fish.

Storks are the subject of numerous myths and legends which have been touched upon in Chap. 7 (Storks and Humans). I have also tried to explore how people have looked at heronries by an examination of writings of naturalists from different periods. Chapter 8 (Conservation) is devoted to conservation. It starts by examining the evidence for the decline of Painted Stork across India and discusses the various conservation problems and their solutions.

In my endeavors to create a smooth text I have tried to remove as many scientific names of plants and animals as possible by grouping them in the Appendix. However, this has not been possible in all cases. Hence, while scientific names of fishes, invertebrates, and some plants have been retained in the text, those of reptiles, birds, and mammals have been placed in the Appendix. A Subject Index has been provided.

**Acknowledgments** During the course of writing this book, when I got thinking about how I became interested in the Painted Stork, I also found myself remembering all those circumstances and people who made a difference. First it was childhood memories of seeing the stork colonies in the Delhi Zoo, in my father's company, not knowing at that tender age what those large white birds sitting on trees were, and ofcourse never imagining in my wildest dreams that one day these birds would occupy such a preeminent position in my life. Years later, visits to the zoo for bird watching became a great way to escape a feeling of claustrophobia, which I often experienced, working in a biological research laboratory at the University of Delhi, while doing my Ph.D. in fish biology.

After much trepidation, one day I made a jump from fish physiology, my original topic of research to field ornithology. Though at one level it was merely switching over from one area of Zoology to another, the process for me was one of re-inventing myself, a type of re-birth. Meanwhile, I passed through several institutions in India and abroad; came in contact with some truly remarkable people and also set about learning a new discipline, behavioral ecology. To all those friends, colleagues and teachers in Delhi, Ahmedabad, England and Germany, who greatly encouraged me in my bird studies, facilitated or helped in innumerable ways, I am grateful. I would particularly like to mention (in alphabetical order of their first names) Bikram Grewal, C.L. Talesara, C.P. Geevan, C.R. Babu, E.P. Jagdeesh, Ed Rispin, Humphrey Sitters, (Late) James Hancock, John D. Goss-Custard, J. Subbarao, Kandarp Kathju, Kartikeya V. Sarabhai, K. Muralidhar, Lalit K. Pande, (Late) Rajmannar, Rebecca Spurk, Richard Caldow, Sangita Gupta, S. Duraiswamy, S.S.S. Sarma, S.V. Goswami, Tejvir Singh, T.R. Rao, Vikas Rai, Zaheer Babar and Zafar Futehally.

Besides the above I must also thank several colleagues who have, over the past 10 years, while I have been a faculty at the University of Delhi, provided help in numerous ways and therefore contributed towards the development of this book; some have hosted me, created opportunities for me to discuss my work by arranging seminars when I visited their institutions; others have given freely of their valuable time by examining reports and dissertations that have emanated from my lab. For all this and more I thank Professor's, Robert Bleweiss (University of Wisconsin, Madison), B.C. Choudhury (WII, Dehradun), Peter Frederick (University of Florida, Gainesville), Anthony J. Gaston (Canadian Wildlife Service), William Karasov (UW, Madison), Jamal Khan (Aligarh Muslim University), Ajith Kumar (NCBS, Bangalore), A.K. Malhotra (Delhi Zoo), R. Nagarajan (Exeter University & AVC College), Asad Rafi Rahmani (BNHS), (Late) Ravi Sankaran (SACON, Coimbatore), G.M. Shah (Kashmir University), Kuldeep Sharma (Jammu University), Anindya Sinha (NIAS, Bangalore) and Lalitha Vijayan (SACON).

My thanks are also due to several (over a dozen) M.Sc. and Ph.D. students of the Department of Environmental Biology who in doing their projects on field ornithology and waterbirds, including several on the Delhi zoo Painted Storks, under my supervision have kept me awake and mostly on my toes. I thank the anonymous referees whose critical questions and sharp comments contributed towards improving the plan of the draft and to Burney's (Tabinda and Khalid Burney) who in London went about hunting, for an old sketch of St. Gile's rookery and eventually found one which I could use (Chapter 2). I thank the various agencies that have funded my work on Painted Stork at different stages: Council for Scientific and Industrial Research (CSIR), India when I started my studies in 1989, and later, after I joined the university faculty, an award of a research scheme funded by the Government of India's Ministry of Environment and Forests (MoEF). This work has also benefited from support from the Waterbird Society, USA and from the University of Delhi's scheme to provide funds to university faculty, for which I am grateful.

***Acknowledgements and credits for material used in this book***

Waterbird Society (USA) for Figures 4.3, 4.4, 4.5, 4.6 originally published in journal Waterbirds (volumes 29 and 32). Prof Trevor Price and Roberts and Co. USA for Figure 1.2. Mital Patel for kindly permitting the use of his original photograph used in Figures 4.2 and 5.5. LYNX EDICIONS, Barcelona, for permitting use of image from del Hoyo et al. (1992). *Handbook of the Birds of the World*. Vol. 1. Ostrich to Ducks. Lynx Edicions, Barcelona in Figure 1.3d. A.R. Rahmani and Abhijit Malekar of BNHS are thanked for kindly making available the photograph by Mrutyumjaya Rao used in Fig. 8.3. Prof. Prajpat Sah of Department of English, Indian Institute of Technology, Kanpur is thanked for translating the Sanskrit verse reproduced in Chapter 7.

For Figures 6.4, 6.5, 6.6, 6.7, originally published in the J. Zoology guidelines available on John Wiley & Sons Ltd. (www.interscience.wiley.com) were followed. For Figures 7.4, 7.5, 7.6, guidelines available on the Indiapost website (http://www.indiapost.gov.in/Philately.html) for reproduction of postage stamps were followed and images accordingly modified.

November, 2010 A.J. Urfi

# References

Love RM (1980) The chemical biology of fishes. Vol 2: advances 1968–1977. Academic Press, London

Urfi AJ, Sen M, Kalam A, Meganathan T (2005) Counting birds in India: a review of methodologies and trends. Curr Sci 89:1997–2003

Urfi AJ (2010) Using heronry birds to monitor urbanization impacts: a case study of Painted Stork *Mycteria leucocephala* nesting in the Delhi Zoo, India. Ambio 39:190–193

# Contents

# Abbreviations

| | |
|---|---|
| AWC | Asian Waterfowl Census |
| BNHS | Bombay Natural History Society |
| CEE | Centre for Environmental Education |
| DI | Dimorphism index |
| DPH | Days post-hatching |
| DZ | Delhi Zoo |
| EE | Environmental Education |
| KDGNP | Keoladeo Ghana National Park |
| LC | Least concern |
| NT | Near threatened |
| SNP | Sultanpur National Park |
| SSD | Sexual size dimorphism |
| VU | Vulnerable |

# Chapter 1
# Introduction

*You can know the name of a bird in all the languages of the world, but when you're finished, you'll know absolutely nothing whatever about the bird... So let's look at the bird and see what it's doing -- that's what counts.*[1]

**Abstract** The Painted Stork is restricted to parts of Asia with a stronghold in India and Sri Lanka. It also occurs in small numbers in Southeast Asia along with its congener the Milky Stork. This introductory chapter outlines various facets of the species covering its nomenclature, systematics, status, broad distribution patterns as well as longevity, dispersal, and instances of hybridization with other species. A comparative account of the genus *Mycteria* is also included. The last section is devoted to a brief discussion about why a study of Painted Stork, as well as other heronry birds, is important in today's conservation context.

## 1.1 What's In a Name?

As an ornithological term, "painted" is not very unique. Many birds' names have this term as a prefix to their English names; "Painted Snipe," for instance. However, while the term is easy to comprehend, there is little in natural history writings to indicate how the Painted Stork came to be so called.

The earliest systematic description of the Painted Stork is by Thomas Pennant who, based on a type specimen collected from (now) Sri Lanka (formerly Ceylon), furnished the first account of the bird in 1790. According to the prevailing trends of ornithological nomenclature, the bird was referred to as "The White-headed Ibis" (*Tantalus leucocephalus*). Its description, reproduced below, was brief, only about 195 words, fairly accurate, yet dropped no hint about how the bird could be visualized as painted. Interestingly, the illustration accompanying the write up in Pennant's book "Indian Zoology" was a highly stylized figure (Fig. 1.1).

[1] Attributed to Prof. Richard Feyman, the Noble prize winning physicist.

A.J. Urfi, *The Painted Stork: Ecology and Conservation*,
DOI 10.1007/978-1-4419-8468-5_1, © Springer Science+Business Media, LLC 2011

**Fig. 1.1** The sketch of Painted Stork (white headed ibis) that appeared in Pennant's book "Indian Zoology" published in 1790. This is possibly the first pictorial description of this species

> In size it is much superior to our largest curlews. The bill is yellow, very long, and thick at the base, and a little incurvated; the nostrils very narrow, and placed near the head; all the fore part of the head is covered with a bare yellow, and seems a continuance of the bill; and the eyes are, in a very singular manner, placed very near its base.
>
> The rest of the head, the neck, back, belly, and secondary feathers, are of a pure white; a transverse broad band of black crosses the breast; the quil – feathers, and coverts of the wings, are black; the coverts of the tail are very long, and of a fine pink color; they hang out and conceal the tail.
>
> The legs and thighs are very long, and of a dull flesh color; the feet semi – palmated, or connected by webs as far as the first joint.
>
> This bird was taken in the isle of *Ceylon,* and kept tame for some time at *Colombo*; it made a snapping noise with its bill like a stork; and, what was remarkable, its fine rosy feathers lost their color during the rainy season.

The Painted Stork, as it is known today, has been referred to by various names in the past such as white-headed ibis (Pennant 1790), pelican ibis (Jerdon 1864; Hume and Oates 1890), etc. It is clear that initially the bird was thought to be a type of ibis and not a true stork. Its bald head and slightly decurved bill probably give the impression of its proximity to ibises. However, the realization soon came about that the bird in question was indeed a stork and the resemblance which it bore to ibises was only superficial. As Blanford (1898) observes,

> This genus [*Pseudotantalus*] and *Tantalus*, which is an American form, chiefly distinguished by its naked neck, were long classed with the Ibises or in a family apart, but they are true storks.

Although storks have traditionally been placed alongside other long-legged waders, recent research however shows that their closest relatives may be New World Vultures (Catharidae) (del Hoyo et al. 1992).

## 1.2 Systematics

Although birds are relatively poorly represented in the fossil record compared to other vertebrates viz. mammals, fishes, and reptiles (Colbert 1969), storks have always been recognized as a well-defined group and their fossil remains known from around the beginning of the Tertiary period. The main radiation of the group is believed to have taken place in the Oligocene period (del Hoyo et al. 1992).

At present 19 species of storks are recognized worldwide, although the taxonomic status of a few is disputed (del Hoyo et al. 1992). Like other taxonomic groups of birds, the stork family too has been subject to several changes, largely due to differences in principle between "lumpers" and "splitters" (Inskipp et al. 1996). Significantly, comparative behavioral studies by Kahl (1971, 1972a–d, 1973, 1974) revealed that Milky, Painted, Yellow-billed, and Wood storks share extremely similar courtship patterns, particularly their up-down displays during courtship. Kahl suggested that all these four species, then divided between the genera *Ibis* and *Mycteria*, should be combined into one genus *Mycteria*.

The genus *Mycteria* was erected by Linnaeus in 1758. The word is derived from the Greek word *mukter*, which means spout or trunk, and Latin *-ius*, which is "resembling" or "connecting." The specific name of Painted Stork derives from the Greek *luekokphalos*, which simply means "white headed" (*leucos*, white and *cephalus*, head). Thus, both the generic and specific names seem justified given that the Painted Stork has a large, heavy trunk like bill and a head which is white, except for the bare skin in adult birds which turns orange in the breeding season (Rahmani in press).

The scientific name of Painted Stork has undergone several changes over the centuries. Pennant (1790) termed it *T. leucocephalus*, while Blanford (1898) referred to it as *Pseudotantalus leucocephalus*. Later ornithologists placed the bird under genus *Ibis*, till finally it was renamed as *Mycteria leucocephalus* (Ali and Ripley 1987). Its specific name was subsequently changed from *leucocephalus* to *leucocephala*, for gender equality (Rahmani in press).

### *1.2.1 Storks Worldwide*

Standard ornithological texts (del Hoyo et al. 1992; Hancock et al. 1992) treat the family Ciconiidae as comprising of three distinct tribes, Mycterini, Ciconiini, and

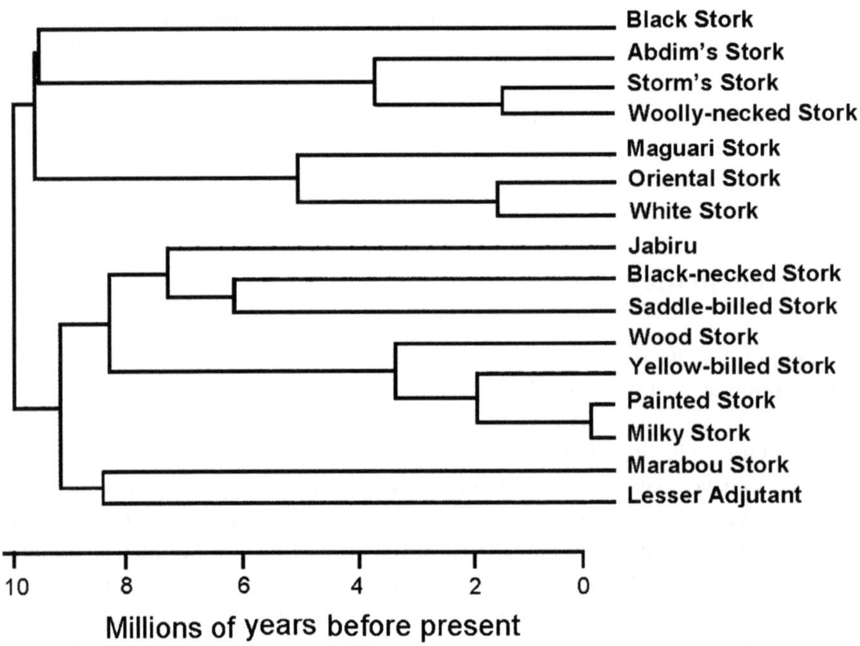


**Fig. 1.2** Dendrogram of stork phylogenies. Note the openbills are not included in this diagram. Redrawn from Price (2008)

Leptoptilini. Mycterini, comprising the wood storks and openbills, includes the genera *Mycteria* (the four species mentioned above) and *Anastomus*, which has two species, Asian Openbill and African Openbill. The openbills, as their name suggests, have an opening or a gap between the mandibles which somehow assists the capture of snails, their principal prey. All the species in this tribe are typified by their smaller size, colonial nesting habit, and a high degree of specialization in their feeding techniques and preying, and consequently in their trophic apparatus. *Mycteria* are all almost exclusively piscivorous and feed by tactolocation. Recent work in molecular systematics (Sibley et al. 1988; Sibley and Ahlquist 1990; Sibley and Monroe 1990; Slikas 1997, 1998) using DNA–DNA hybridization reveals that the Painted Stork and Milky Stork are phylogenetically very close and the two species could have separated only recently in evolutionary time (Fig. 1.2). In fact, as we shall discuss at other places in this book, the two species co-occur in some parts of Southeast Asia and even interbreed at some sites, both in the wild as well as in captive conditions.

Ciconiini, comprising the "typical storks," includes storks belonging to genus *Ciconia*, many of which are true migrants covering long distances across the globe and having a mixed diet which includes insects, terrestrial arthropods as well as reptiles, amphibians, and fishes. The Asian species of interest in this tribe include the Black Stork and Woolly-necked Stork. The latter is a resident all across India and parts of SE Asia and Africa. Other than these is of course the European White Stork which is the subject of innumerable myths and tales in its breeding range not only in Europe but also in Africa and Asia (details in Chap. 7), which constitutes its wintering range. Other storks in this tribe are Abdim's Stork, restricted to Africa; Storm's Stork, restricted to certain parts of SE Asia; Maguari Stork, restricted to South America; and the Oriental White Stork in East Asia.

Tribe Leptoptilini comprising "giant storks" basically constitutes two forms, genus *Ephippiorhynchus* and *Jabiru* in one group and genus *Leptoptilos* in the other. In the former are included Black-necked Stork – of which there is an Asian race as well as an Australian race – the Saddlebill resident of tropical Africa, the Jabiru in the neotropical region. The second group in this tribe includes the adjutant storks (Lesser Adjutant and Greater Adjutant) which can cover long distances on the wing. Both species are found in Asia and are highly endangered (Birdlife International 2001). Their African counterpart is the Marabou. Another stork The Shoebill is normally placed in a monospecific family, and sometimes in its own order (Balaenicipitiformes) since its systematics is unclear.

## 1.3 Hybridization

Although the Painted Stork is believed to be declining throughout its range (Birdlife International 2010), there are two crucial ways in which it manifests itself as a problem. First, in some places, for example in Japan it is regarded as an invasive species and has colonized waterways and wetlands (Eguchi and Amano 2004). Second, in SE Asia where it co-occurs with its congener the Milky Stork, which is evaluated as vulnerable (Birdlife International 2001) and therefore conservationally even more precariously placed, it hybridizes easily. This is a matter of concern from the point of view of developing reintroduction programs for the Milky Stork because its wild gene pool may be negatively impacted by hybridization with the Painted Stork.

Several cases of hybridization between Painted Stork and other waterbirds, from the wild, are on record. At Ang Trapeang Thmor Sarus Crane Reserve in Cambodia where Milky Stork and Painted Stork coexist, hybridization between the two has been reported (Eames 2007). This is among the few cases of hybridization in the wild. At Zoo Negra in Kuala Lumpur, Malaysia, instances of hybridization between Painted Stork, Lesser Adjutant and Milky Stork have also been recorded (Li et al. 2006). Since Negra Zoo colonies, much like Delhi Zoo heronries, consist of free range birds, this too is a case of hybridization in wild or natural conditions.

Under captivity conditions, several instances of hybridization between Painted Stork and other species are known. Painted Storks have been recorded to cross with Lesser Adjutant in Dehiwela Zoo, Colombo (Henry 1971) and with Milky Stork in Dusit Zoo, Bangkok, Thailand and also from Singapore Zoo, Singapore (Li et al. 2006). Cases of hybridization between Painted Stork and other birds such as Spoonbill, Black-headed Ibis are also known from zoos across India.

## 1.4 *Mycteria*: A Comparative Study

Jerdon (1864) noted that the Pelican Ibis (Painted Stork)

> … is replaced in the Malay countries by T. lacteus, Temminck ; and there are other species in Africa and America, T. ibis, and T. loculator.

This statement was made more than one and a half century ago and therefore, the idea that these four species, now grouped under one genus, constituted a closely related group, though flung across three different continents, has been around for some time. The same point has been reiterated by Blanford (1898). For a better understanding of the biology of Painted Stork, a comparative study of *Mycteria* is therefore in order.

Morphologically, the four species of *Mycteria* are distinct and there can be little cause for confusion. The Wood Stork is a comparatively smaller bird (Table 1.1) with a white body. Its head and neck are black and unfeathered and there are bony plates on the cap (marked by an arrow, Fig. 1.3). Its legs are also black. The other three *Mycteria* are all different from the Wood Stork but quite similar to each other. In all three the bill is waxy yellow and the legs pale red or flesh colored. They have a patch of skin, deeper yellow, or orange colored, extending from the base of the bill upto the eyes and a little beyond. The size of the unfeathered parts of the head, the so-called "bald head" so to speak, varies seasonally. For instance in the Painted Stork, it becomes very pronounced in the breeding season. In the case of the other two species, it appears in sharper contrast against the bill (marked in Fig. 1.3).

In all of these three species, the body plumage is white with black, in varying degrees, especially on the primaries. The Milky Stork is wholly white with a rim of black feathers on the wings (Fig. 1.3). The Yellow-billed and Painted Stork have more patches of black, besides pink feathers, which are lacking in the Milky Stork. In the case of the Yellow-billed Stork, the pink is more pronounced on the back and shoulders; but in the case of the Painted Stork, it is only restricted to the tail region and appears only in the breeding season. Morphologically, these two species also differ in another important respect. The Painted Stork has a broad pectoral band (Fig. 1.3) which is absent in the Yellow-billed Stork.

The geographical distribution of each of the four species is different. The Wood Stork occurs in some of the southeastern states of the USA, Mexico, through Central and South America to northern part of Argentina. Throughout its range, it inhabits a range of wetlands including mangroves, gallery forest, damp grassland, freshwater marshes and swamps, etc. The Yellow-billed Stork is restricted to the African continent, south of Sahara, and Madagascar. However, it straggles into Palearctic Africa in Morocco, Tunisia, and Egypt. The other two species of *Mycteria* are both inhabitants of Asia. While the Painted Stork has a stronghold in the Indian subcontinent, its range spreads eastward up till Indo-China. The Milky Stork occurs in South Vietnam, Kampuchea, Peninsular Malaysia, and the Indonesian islands of Sumatra, Java, and Sulawesi. In Cambodia, Malaysia, and Vietnam the two species co-occur, but in Indonesia the Painted Stork is absent. More details about the distribution of Painted Stork are given in the subsequent section.

**Table 1.1** Synopsis of the four species of *Mycteria*

| Parameters | Species: *M. americana* | *M. ibis* | *M. leucocephala* | *M. cinerea* |
|---|---|---|---|---|
| English name (s) | Wood Stork/Wood Ibis | Yellow-billed Stork | Painted Stork | Milky Stork |
| French | Tantale d'Am rique | Tantale ibis | Tantale indien | Tantale blanc |
| German | Waldstorch | Nimmersatt | Buntstorch | Milchstorch |
| Spanish | Tántalo Americano | Tántalo Africano | Tántalo Indio | Tántalo Malayo |
| Taxonomy | *Mycteria americana* Linnaeus 1758, Brazil *ex* Marcgraf | *Tantalus ibis* Linnaeus, 1766, Egypt | *Tantalus leucocephalus* Pennant, 1769, Ceylon | *Tantalus cinereus* Raffles, 1822, Sumatra. |
| Nesting | Colonial tree nester, sometimes in mixed colonies | Colonial tree nester often with other species | Colonial tree nester often with other species | Colonial tree nester |
| Clutch | Average 3 (2–5) | 2–3, rarely 4 | Normally 3–4 (2–5) | 1–4 |
| Incubation | 28–32 days | Prob ca. 30 days | ca. 30 days | – |
| Fledging | 60–65 days | ca. 55 days | ca. 60 days | – |
| Sexual maturity | Four years though nesting normally not successful until fifth year | Probably a minimum of 3 years | 4 years | – |
| Conservation status | LC | LC | NT | VU |
| Body weight (range) (kg) | 2.05–3.30 | 1.19–2.33 | 2.0–3.5 | – |
| Tarsus (range) (mm) | 175–208 | 195–230 | 194–250 | 188–225 |
| Egg wt (range) (g) | 80–83 | 62–77 | 66–98 | 77 |
| Longevity captivity (c)/ wild (w) years | 30+ (c)/11 years, 8 months (w) | 19+ (c) | 28+ (c) | – |

*Sources*: Chiefly del Hoyo et al. (1992) and Hancock et al. (1992)

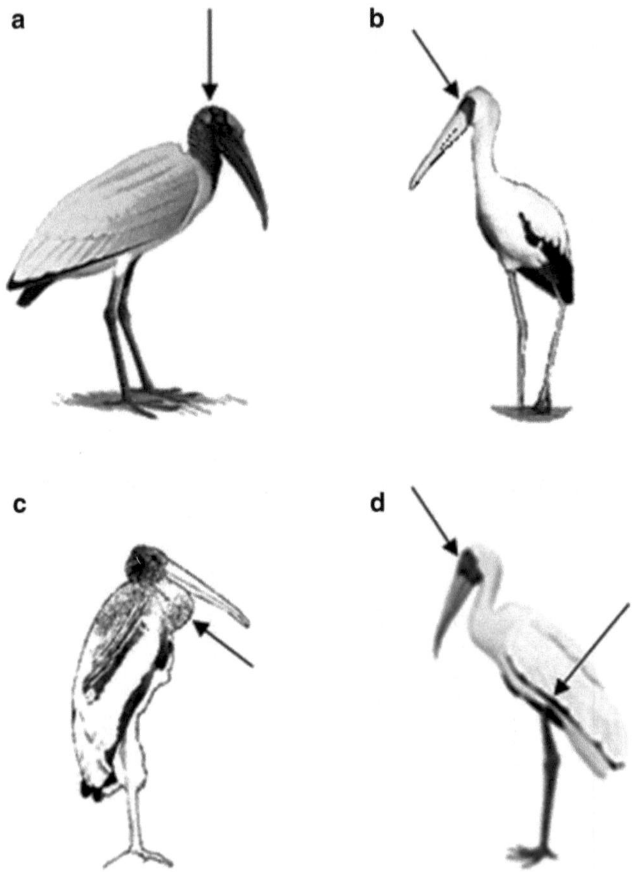


**Fig. 1.3** The basic external features of the four species of *Mycteria*. (**a**) Wood Stork, (**b**) Yellow-billed Stork, (**c**) Painted Stork, and (**d**) Milky Stork. See text for explanation of *arrows*

## 1.5 Status and Distribution

An important source of information regarding the status and distribution of waterbirds across Asia emanates from the Asian Waterfowl Census (AWC). This program, initiated by International Waterfowl & Wetlands Research Bureau (IWRB), Slimbridge, in 1987 and deriving its inputs largely through volunteer efforts has played a crucial role in terms of enhancing our understanding about waterfowl distribution patterns in the region (Urfi et al. 2005). However, since the AWC data also has certain limitations, it should be interpreted with care. The census is mostly carried out at well-known, important wetlands in the month of January. Painted Storks are in fact equally likely to occur at smaller wetlands, flooded agricultural

**Table 1.2** Select species of large colonially nesting waders and their maximum numbers reported for the whole region under the coverage of AWC (including Asia and SE Asia)

| | Years | | | |
|---|---|---|---|---|
| Species | 1987–1992 | 1993–1997 | 1998–2002 | 2003–2007 |
| Milky Stork | 432 | 520 | 133 | 43 |
| Painted Stork | 5,250 | 6,009 | 6,648 | 9,402 |
| Asian Openbill | 21,030 | 24,425 | 63,261 | 258,105 |
| Black-headed Ibis | 8,208 | 7,185 | 4,833 | 5,907 |

The data covers the period starting from the inception of the program in 1987 till 2007
*Source*: Li et al. (2009)

fields, small ponds and marshes, etc., lying outside the coverage of the AWC. Hence, it is likely to be grossly underreported in bird counts (Sundar 2006; Urfi et al. 2005). Also, the participation of volunteers as well as the number of sites covered in the AWC is not constant from year to year and for this reason the number of birds reported is entirely dependant upon effort. (In fact, as would be expected, the number of Painted Stork reported and sites covered/volunteers involved are correlated.) However, what comes out unambiguously from the AWC is that the Painted Stork is largely a bird of the Indian subcontinent although it also occurs in smaller numbers in some countries of SE Asia. Second, India and Sri Lanka are two countries which are most important from the point of view of this stork since in these countries sizeable numbers have been reported in most years. Pakistan also figures as a country of some importance even though for some years the Painted Storks were not recorded from here at all. Tonle Sap in Cambodia is perhaps the most significant site for Painted Stork in SE Asia.

Although the Painted Stork is considered one of the most numerous and secure of Asian storks (del Hoyo et al. 1992), the increasing impact of habitat loss, disturbance, pollution, hunting, and several other factors are believed to be causing a decline in its populations. According to some estimates there are approximately 15,000 individuals in South Asia and fewer than 10,000 in Southeast Asia, with populations declining throughout (Scott and Rose 1989).

It is interesting to see how the Painted Stork compares with some other large-sized colonially nesting waders, particularly those which coexist with it at mixed species heronries. These species are principally the Asian Openbill and Black-headed Ibis. As shown in Table 1.2, Asian Openbill is the most abundant of the three, with nearly 260,000 individuals having been reported in some years during 2003–2007. Thus, there are 26 times more openbills throughout Asia than Painted Storks. The distribution range of Asian Openbill is also very wide in this region, stretching from India to SE Asia (Li et al. 2009). The Black-headed Ibis is recorded in fairly large numbers too, nearly as much as the Painted Stork. If the Milky Stork is also included in the picture, then it comes out extremely poorly in terms of population size (Table 1.2), besides the fact that its distribution is also very restricted.

### 1.5.1 Distribution in the Indian Subcontinent

Generally speaking, food availability is the single most important factor governing different aspects of ecology, distribution, and biology of storks. Storks are good examples of birds that are "reproductively limited" by food (Lack 1968), and it is in this context that we must view the distribution of the Painted Stork across its range. For those birds which are specialized foragers, certain conditions with respect to food availability, abundance, and food type must be met with if they are to breed successfully in any given region. Thus, mere presence of an individual in any locality does not necessarily imply fitness for breeding, which is heavily dependant upon food cycles (del Hoyo et al. 1992). In many regional works such as Ali and Ripley (1987), Kazmierczak (2000), and Grimmett et al. (1998) as well as in specialized texts such as del Hoyo et al. (1992) and Hancock et al. (1992), the distribution of the Painted Stork is depicted as being spread across a very wide area, extending from Pakistan in the West to areas in Indo-China in SE Asia. Grimmett et al. (1998) and Rasmussen and Anderton (2005) have provided updated distribution maps in their works, showing a more restricted distribution of the species.

From Pakistan, the Painted Stork is mainly recorded along the river Indus and in the major lakes in Sind (Hasan 2001). It is spotted in Punjab along the larger river channels in the plains. Pyhälä (2001) considers the Painted Stork a vagrant in Pakistan but lists several sites from where it has been sighted in recent times. These include Rawal lake (three seen on 6 June 1980 and a single bird seen in the same year on 30 June). During 1992–1996 Painted Storks were recorded in winter surveys in Thatta and Badin. Among the interesting observations on Painted Stork from Pakistan is that of 20 birds seen roosting on *Ficus* trees by Balloki headworks on the river Ravi (Roberts 1991). Given the time of year, as mentioned in the above report, it would seem that the birds were in preparation for nesting. But, regarding nesting most authors from Pakistan are ambiguous and no specific site seems to be on record from recent times. While Roberts (1991) mentions that the bird breeds in the Indus Delta, he does not furnish any site details. Pyhälä (2001) provides a picture of Painted Storks at nest but stops short of mentioning any specific nesting site.

Among some very old nesting records of Painted Stork from areas what were in former times British India but lie now in Pakistan is an account of a colony from E. Narra in Sind. According to a correspondent in Hume and Oates (1890),

> I found a large colony of these birds breeding in the end of February. The nests, which seemed very small for the size of the bird, were rude stick platforms built on decayed trees about 6 or 8 feet over water level. The nests all contained young birds very many nearly able to fly.

East of the river Indus, close to the India–Pakistan border, there appears to be a zone from where no Painted Stork sightings have been reported (Rasmussen and Anderton 2005), probably due to the presence of the Indian desert (The Thar desert). It is only when one reaches southern and eastern parts of Rajasthan that Painted Stork become abundant. However, what appears to be a gap (Fig. 1.4) on the western fringe of the distribution of the Painted Stork may actually not be so. First, probably few birdwatchers frequent these areas due to the desert and hence an underreporting

**Fig. 1.4** Distribution map of Painted Stork from the Indian subcontinent. The *shaded areas* show the distribution. *A & N is* Andaman and Nicobar islands, part of Indian territory. *Sources*: Rasmussen and Anderton (2005), Li et al. (2009), and Grimmett et al. (1998)

is to be expected. But more importantly, environmental changes are being brought about in the Thar desert due to the construction of a major irrigation canal known as "Indira Gandhi Nahar Pariyojna," which runs quite close to the India–Pakistan border and passes close to the Indian city Jodhpur (Rahmani 1997). Reports suggest that the local ecology and landscape are changing drastically due to the irrigation canal and the number of waterfowl has increased (Sivaperuman et al. 2009).

From the northern part of the Indian subcontinent, in the Indian Punjab (as well as parts of Pakistani Punjab) the Painted Stork is well represented. In fact at a site called Lehalan 600 were reported in the AWC of 1990 (Li et al. 2009). This number is above the 1% (=250 for South Asia) of the total population estimate, as per criteria 6 of the Ramsar Convention. Painted Storks have also been recorded, albeit in small numbers, from several sites in the hill state of North India, Uttarakhand. These are Asan Barrage (Dehradun), Corbet Tiger Reserve (Pauri Garhwal and Nainital districts), and Rajaji National Park (Dehradun, Haridwar, and Garhwal) (Islam and Rahmani 2004). All of these sites are in the Himalayan foothills and not high up in the mountains.

In Nepal the Painted Stork is recorded as a "rare summer visitor and resident below 250 m" (Baral and Inskipp 2005). The sites where it has been recorded in recent years are mostly from the southern parts of the country and lie in the *terai* or the foothills of the Himalayan Mountains. These include Barandabhar forests and wetlands, Royal Bardia National park, Royal Chitwan National park, Koshi Tappu Wildlife Reserve and Koshi Barrage, farmlands in Lumbini areas, and Royal Sukla

Phanta Wildlife reserve. No specific nesting records are known from these sites or indeed from anywhere else in Nepal. As far as China is concerned, the species was previously reported as a common summer visitor in the south, probably breeding, but now it is rare and possibly extinct (Birdlife International 2001, 2010).

In India, the Painted Storks occur across the length and breadth of the country barring some parts. For instance, in Kerala, in the extreme south and in Northeastern India it is a rare visitor. In Assam, Painted Stork is rare and sighted sporadically, mostly from the Bhramaputra river valley (Choudhury 2000). In Bangladesh the bird is not common (Khan 1987) while in Myanmar it was formerly resident in the central region and a visitor throughout. According to Smythies (1940) it was "a very common bird on the coastal mudflats, seen both solitarily as well as in flocks," also frequenting jheels, flooded paddy fields, and marshes. However, the current status of the Painted Stork in Myanmar is largely unknown but clearly it is a very rare bird here (Birdlife International 2001, 2010). In the AWC only four individuals were recorded in 1992 suggesting that its occurrence here is highly sporadic. There is no record of Painted Stork from Andaman and Nicobar Islands in the Bay of Bengal (Fig. 1.4).

Sri Lanka is of course the second most important country for Painted Stork. Here, it is locally abundant, particularly in the dry zone (del Hoyo et al. 1992; Grimmett et al. 1998). It is common in the northern half of the island and down the eastern low country to about the Tangalla district, wherever tanks, large swamps, and secluded lagoons give it safe feeding grounds (Henry 1971). Among some very old nesting records is that of Uduwila (sic) Tank (Hume and Oates 1890). In modern distribution maps, the species is shown to be seemingly absent from the center of the island country (Fig. 1.4). Large nesting colonies have been recorded from Kumana-Villu of the Yala East National Park (Kaluthota et al. 2005). Painted Storks also occur in large numbers from other protected areas of Sri Lanka such as Ruhina National Park (Santiapillai et al. 2001).

## *1.5.2 Distribution in Southeast Asia*

In Indo-China the Painted Stork is reported from several countries (Fig. 1.5). From a global perspective, it would seem as if the distribution of Painted Stork was discontinuous but most probably this is only because in large parts of this region the bird has been persecuted or its habitat has come under threat and hence large gaps are to be seen. Thus, this is not a case of true discontinuous distribution in a biogeographical sense (Ripley and Beehler 1990).

In Laos, Painted Stork was previously widespread but now it is rare. In Vietnam also it was formerly a widespread resident but now it is a rare nonbreeding visitor (Grimmett et al. 1998; Robson 2000). In Malaysia it was previously regular but is now a vagrant (Wells 1999; Robson 2000) where it frequents freshwater marshes, lakes and reservoirs, flooded fields, rice paddies, freshwater swamp forest, riverbanks, intertidal mudflats, and saltpans. According to Delacour (1947) it was generally found in the north Malay States. It is only in Cambodia that the Painted

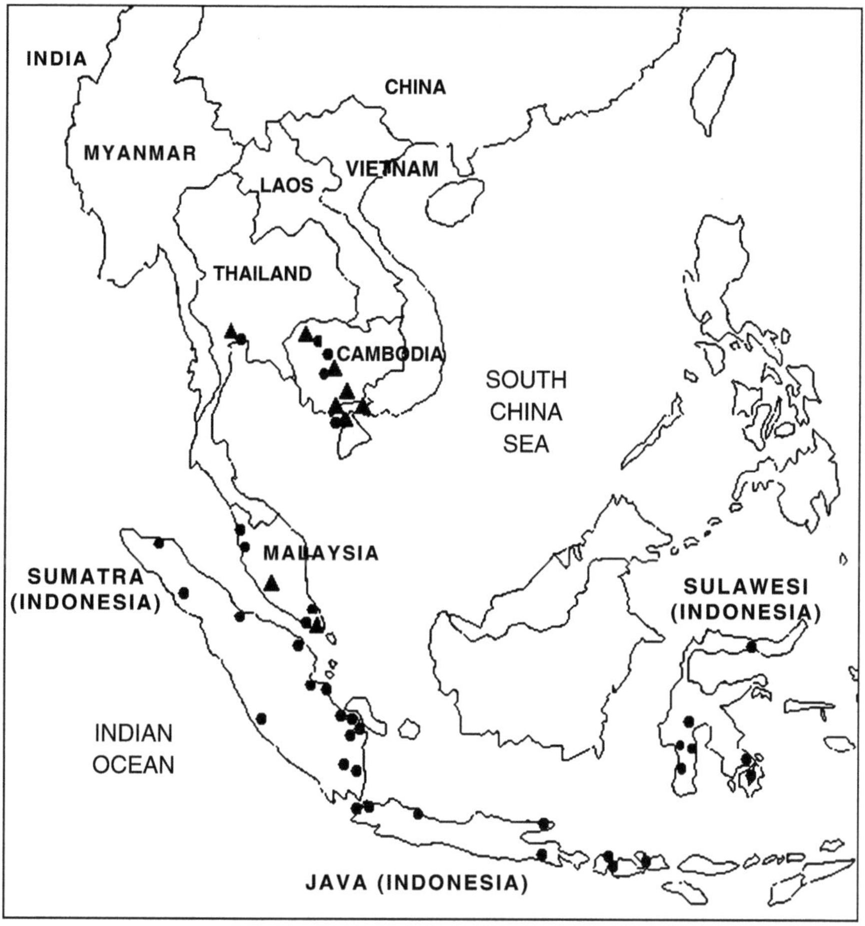


**Fig. 1.5** Map to show some of the sites in SE Asia from where Painted Stork (*filled triangles*) and Milky Stork (*filled circles*) have been recorded. Based largely on Li et al. (2006)

Stork seems to be doing well. Here it is a local resident, with a minimum of several hundred pairs breeding at Tonle Sap (Mundkur et al. 1995; Robson 2000), which is situated within the floodplain of the Mekong River and is the largest natural freshwater lake in Southeast Asia. The Tonle Sap Great Lake colony is probably the largest in SE Asia (Campbell et al. 2006) and an estimated ca. 20% of the regional Painted Stork population exists here. The site is also important as the only freshwater colony of Milky Stork in the world. Besides Tonle Sap a small colony of Painted Storks has been reported at Ang Trapeang Thmor Sarus Crane, where hybridization with Milky Stork has also been suspected (Eames 2007). Finally, in Thailand the Painted Stork was previously regarded as a common breeder in the south but now it is on the verge of extinction, and is recorded sporadically in small numbers (Wells 1999; Robson 2000).

## 1.6 Basic Information About the Painted Stork

If one were to list all the basic facts about the Painted Stork, in an encyclopedic fashion, then it might fill a full chapter or even a book. It may suffice here to say that the basic biological parameters pertaining to this species, including morphology, anatomy, ecology, etc. have been dealt with in standard texts. Information on nearly all aspects including its hematology is available (Salakij et al. 2003) though scattered. In this section only four interesting aspects of Painted Stork are discussed below.

### *1.6.1 Longevity*

In a review of longevity data on various species of storks in captivity (Brouwer et al. 1992), only two records pertain to Painted Stork. One was a living specimen in Cairo Zoo which was known to be 10 years old in 1924. Another is of a male bird in Amsterdam Zoo which lived for a period of 28 years, 7 months, and 17 days, till it died on 9 April 1914.

If we plot longevity data of different species of storks (Brouwer et al. 1992) against their body weights (del Hoyo et al. 1992; Hancock et al. 1992) then, as expected, a positive relationship emerges with Painted Stork placed in the middle (Fig. 1.6). Being large storks, though not the largest, their longevity is higher than smaller-sized storks. However, it has to be borne in mind that longevity in captive conditions is always more than under natural conditions since the birds live in a highly protected environment and there is absence of predators and disease.

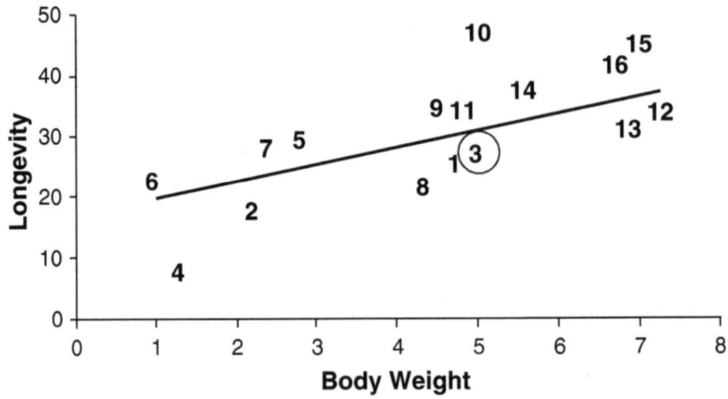


**Fig. 1.6** Graph to show the relationship between body weight (kg) and longevity (years) under captive conditions for some species of storks. The value of Painted Stork is *encircled*. The numbers represent different species. *1* Wood Stork; *2* Yellow-billed Stork; *3* Painted Stork; *4* African Openbill; *5* Black Stork; *6* Abdim's Stork; *7* Woolly-necked Stork; *8* Maguari Stork; *9* White Stork; *10* Oriental Stork; *11* Black-necked Stork; *12* Saddle-billed Stork; *13* Jabiru; *14* Lesser Adjutant; *15* Marabou Stork; *16* Shoebill. *Sources*: Brouwer et al. (1992), del Hoyo et al. (1992), and Hancock et al. (1992). For other details see text

### 1.6.2 *Dispersal*

Unlike the White Stork which is a long distance migrant, and the adjutant storks which are also known to disperse over fairly long distances, Painted Storks are resident birds and do not exhibit true migration. During much of their nonbreeding phase, Painted Storks wander across large areas looking for food and habitat (Ali and Ripley 1987). Although, the movements of the Painted Stork would be expected to be on a subcontinental scale but due to paucity of data from ringed birds it is difficult to say exactly how far they travel. Its congener Wood Stork is relatively philopatric, i.e., by comparison with other wading birds in Southeastern US (Frederick and Meyer 2008) on this basis it would be reasonable to assume that so are the Painted Stork.

As a part of the BNHSs project to study the movement of Indian birds about 500–600 storks, cormorant and other colonial birds were ringed at Keoladeo, Bharatpur. Only four recoveries of Openbill are on record, ranging from 180 to 510 miles from Bharatpur. Interestingly, all are from areas lying east of Bharatpur (Ali 1959). While of course saying nothing about the Painted Stork, this study tells us something about the range across which other species of Indian storks, nesting in the same colonies as the Painted Stork, disperse.

### 1.6.3 *Known Enemies*

Being fairly large birds, it would be expected that Painted Stork will not have any major biological threats as adults save for disease and predation as eggs or nestlings (dealt with in detail in Chap. 4). However, instances of Painted Stork being attacked by raptors, trained or untrained, are on record. For instance Jerdon (1862) mentions that the Peregrine Falcon and Shahin Falcon, both popular birds in falconry, can be trained to attack large birds like Painted Stork. Jerdon adds

> Nisaetus Bonelli [sic] preys by preference on various kinds of game, hares… and according to the testimony of Shikarees, it has been known to strike down the douk

Most likely such cases are the exception rather than the norm. The biggest threat which the Painted Stork faces today is from man and his activities which adversely affect the natural environment.

### 1.6.4 *Local Names*

Across much of the Hindi speaking belt of Northern India, the Painted Stork is called Jănghil, Dhōk (Ali and Ripley 1987). The Painted Stork has a name in almost all the different languages spoken across South India (Table 1.3). Even in those places which lie in the peripheral parts of the Painted Stork's range or in those areas

**Table 1.3** Names of Painted Stork in different languages spoken across its range

| Region | Language | Local names |
|---|---|---|
| North Indian states | Hindi | Jănghil/Jaunghal/Jaunghil/Jangkil, Dokh/ Dhōk |
| North India | Mirshikars dialect | Kăkāri |
| Western India/Gujarat | Kutch, Gujarati | Chitrōda/Chitrodi Bagga |
| Pakistan | Sindhi | Lungduk |
| Eastern India | Bengali | Kat saranga/Kat-sarunga/Ram-jhankar/ Sona-janga/Sona jăngha, Janghir/Jănghil, Rām Jhănkar |
| South India | Telugu | Yerri kali-konga/Yeru Kălā Konga |
| South India | Kannada | Banada Kokkare |
| South India | Tamil | Singa Nareh/Chĕngă Nārăi |
| Sri Lanka | Tamil (of Sri Lanka) | Singanareh/Changaveelanary/Sănguvălai Nārăi |
| Sri Lanka | Sinhala | Das Tuduwa |
| Myanmar | Burmese | Datuduwa, Cing., Hnet-kya |
| Malaysia | Malay | Burong Upeh |

*Sources*: Jerdon (1864), Finn (1906), Glenister (1971), Ali and Ripley (1987), and Manu and Jolly (2000)

where it is not such a common or abundant bird, local names exist. For instance in "Bangla," the language spoken in the Indian state West Bengal and the neighboring country Bangladesh, as well as Burmese spoken in Myanmar, several names for the Painted Stork have been recorded (Table 1.3). In these places the Painted Stork is not a very common bird. In Malaya, the Painted Stork is known as *burong upiah*, which is also the name for the Milky Stork. "Burong" means "bird" in Malay, and may be affixed to many bird names (Glenister 1971).

## 1.7 The Significance of Studying Painted Stork and Similar Birds

There are several good reasons why a study of Painted Stork can be both meaningful and fruitful. A large, carnivorous (fish eating) bird is always a good model for ecological studies simply because it can be easily seen and counted. The Painted Stork is an indicator of its habitat, wetlands, which merit attention on their own due to their endangered status. Across much of Asia, the Indian subcontinent particularly, wetlands are being lost to land encroachments, pollution, and other factors. Yet, several crucial aspects of their ecology, particularly the role of birds in their functioning remains largely unexplored.

Since environmental toxins, particularly organochlorine pesticides, travel fast along aquatic food chains, by bio-magnification, piscivorous birds like the Painted Stork exhibit early signals of contamination and therefore their study assumes a special significance.

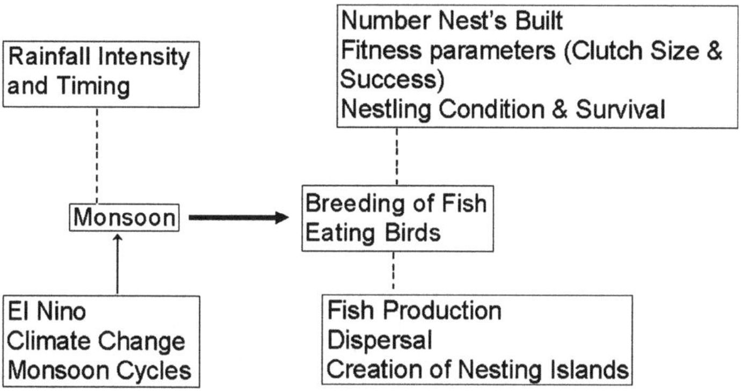


**Fig. 1.7** A scheme to show the linkages between the monsoon and Painted Stork reproduction. This scheme forms the basis of developing a conservation monitoring program for fish-eating birds, discussed in the text

The Painted Stork is a colonial waterbird and it provides us an opportunity to explore questions about avian coloniality, a topic which has fascinated ornithologists and evolutionary biologists for a long time. While the literature on this subject is extensive, detailed studies on nonpasserines and particularly "heronry birds," i.e., stork, ibis, spoonbill, egret, and herons, are few in number and it is here that studies on birds like the Painted Stork can yield valuable insights and understanding.

The food supply of fish-eating birds like Painted Stork is strongly influenced by the monsoon, this adds another dimension to their study. Much of Asia has a predominantly monsoonal climate in which there is a dry phase and a wet phase, in which wetlands are revived and aquifers recharged. In spite of fears that due to climate change the monsoon regime is going to be affected, there is yet no good model for studying the effects of seasonal rains on biodiversity of wetlands.

Figure 1.7 shows the broad connections that constitute the theme of this book. Being a fish-eating bird, Painted Stork reproduction is closely linked to the performance of the monsoon, which is the primary driver in the sense of triggering off their food cycles. In various sections of this book, an attempt is made to unravel the linkages between the monsoon rains, nesting of Painted Stork, and how monitoring fitness parameters can help us to understand the underlying ecological processes better.

While wetlands – the foraging habitat of Painted Stork – are gravely threatened, equally vulnerable today are the nesting colonies, "heronries," of such birds. India still has many large heronries. Of these, over 50% according to some estimates (Subramanya 1996) are located in urban premises. The Delhi Zoo colony, which is no less than a bird sanctuary, is actually not so unique after all. Indeed in many other places, colonial waterbirds nest in large numbers in urban parks and gardens. Perhaps, due to the fact that their natural nesting areas are getting fewer in the countryside, due to cutting down trees and habitat loss, inside urban premises, colonial waterbirds are able to find suitable nesting sites and safety from prowlers and miscreants. Be that as it may, these birds are also indicators of the changes, of a different kind going on in their environment, namely the spread of urbanization and the increasing cover of built-up areas. Monitoring these birds is akin to keeping tabs on the environment and can therefore yield interesting insights for city planning and development (Urfi 2010).

## References

Ali S (1959) Local movements of resident birds. J Bombay Nat Hist Soc 56:346–347

Ali S, Ripley SD (1987) Handbook of the birds of India and Pakistan together with those of Bangladesh, Nepal, Bhutan and Sri Lanka. Oxford University Press, Delhi

Baral HS, Inskipp C (2005) Important bird areas in Nepal: key sites for conservation. Birdconservation Nepal and Bird Life International, Kathmandu, Nepal

BirdLife International (2001) Threatened birds of Asia: the birdlife international red data book. BirdLife International, Cambridge

BirdLife International (2010) Species factsheet: *Mycteria leucocephala*. http://www.birdlife.org. Accessed 14 June 2010

Blanford WT (1898) The fauna of British India, including Ceylon and Burma. Birds, vol IV. Taylor and Francis, London

Brouwer K, Jones ML, King CE, Schifter H (1992) Longevity and breeding re cords of storks *Ciconiidae* in captivity. Int Zoo Yearbook 31:131–139

Campbell IC, Poole C, Giesen W, Valbo-Jorgensen J (2006) Species diversity and ecology of Tonle Sap Great Lake, Cambodia. Aquat Sci 68:355–373

Choudhury A (2000) The birds of Assam. Gibbon Books, Gawhati

Colbert EH (1969) Evolution of the vertebrates. Wiley, New York

del Hoyo J, Elliott A, Sargatal J (eds) (1992) Handbook of the birds of the world, vol 1. Lynx Edicions, Barcelona

Delacour J (1947) Birds of Malaysia. Macmillan Company, New York

Eames JC (2007) Apparent hybridization of Milky and Painted Storks at Ang Trapaeng Thmor Sarus Crane conservation area, Cambodia. Babbler 22&23:21

Eguchi K, Amano HE (2004) Invasive birds in Japan. Glob Environ Res 8:29–39

Finn F (1906) How to known the Indian waders. Thacker, Spink & Co., Calcutta

Frederick P, Meyer KD (2008) Longevity and size of wood stork (*Mycteria americana*) colonies in Florida as guides for an effective monitoring strategy in the southeastern United States. Waterbirds 31:12–18

Glenister AG (1971) The birds of the Malay Peninsula, Singapore and Penang. Oxford University Press, Singapore

Grimmett R, Inskipp C, Inskipp T (1998) Birds of the Indian subcontinent. Oxford University Press, New Delhi

Hancock JA, Kushlan JA, Kahl MP (1992) Storks, ibises and spoonbills of the World. Academic, London

Hasan M (2001) Birds of the Indus. Oxford University Press, Karachi

Henry GM (1971) A guide to the birds of Ceylon, 2nd edn. K.V. De Silva and Sons, Kandy, Sri Lanka

Hume AO, Oates EW (1890) The nests and eggs of Indian birds, vol III. Taylor & Francis, London

Inskipp T, Lindsey N, Duckworth W (1996) An annotated checklist of the birds of the oriental region. Oriental Bird Club, England

Islam MZ, Rahmani AR (2004) Important bird areas in India: priority sites for conservation. Bombay Natural History Society, Mumbai

Jerdon TC (1862) The birds of India, vol I. Military Orphan Press, Calcutta

Jerdon TC (1864) The birds of India, vol III. George Wyman & Co., Calcutta

Kahl MP (1971) The courtship of storks. Nat History 80:36–45

Kahl MP (1972a) Comparative ethology of the Ciconiidae. The Wood Storks (genera *Mycteria* and *Ibis*). Ibis 114:15–29
Kahl MP (1972b) Comparative ethology of the Ciconiidae. Part 4. The "Typical" Storks (genera *Ciconia, Sphenorhynchus, Dissoura,* and *Euxenura*). Z Tierpsychol 30:225–252
Kahl MP (1972c) Comparative ethology of the Ciconiidae. Part 5. The Openbill Storks (genus *Anastomus*). J Ornithol 113:121–137
Kahl MP (1972d) A revision of the Family Ciconiidae (Aves). J Zool (Lond) 167:451–461
Kahl MP (1973) Comparative ethology of the Ciconiidae. Part 6. The Blacknecked, Saddlebill and Jabiru storks (genera *Xenorhynchus, Ephippiorhynchus* and *Jabiru*). Condor 75:17–27
Kahl MP (1974) Comparative behavior and ecology of Asian storks. In: National geographic society research reports, 1967 projects. National Geographic Society, Washington
Kaluthota CD, Gamage SN, Kaluthota ULS (2005) Breeding status of the Painted Stork *Mycteria leucocephala* in the Kumana-Villu of the Yala East National Park Forestry and Environment Symposium, Colombo, Sri Lanka
Kazmierczak K (2000) A field guide to the birds of India, Sri Lanka, Pakistan, Nepal, Bhutan, Bangladesh and the Maldives. Om Book Service, New Delhi
Khan MAR (1987) Conservation of storks and other waterbirds in Bangladesh. Colon Waterbirds 10:229–235
Lack D (1968) Ecological adaptations for breeding in birds. Chapman and Hall, London
Li ZWD, Yatim SH, Howes J, Ilias R (2006) Status overview and recommendations for the conservation of Milky Stork Mycteria cinerea in Malaysia: final report of the 2004/2006 Milky Stork field surveys in the Matang Mangrove Forest, Perak, Malaysia. Wetlands International and the Department of Wildlife and National Parks, Peninsular Malaysia, Kuala Lumpur, Malaysia
Li ZWD, Bloem A, Delany S, Martakis G, Quintero JO (2009) Status of waterbirds in Asia. Results of the Asian waterbird census: 1987–2007. Wetlands International, Kuala Lumpur, Malaysia
Manu K, Jolly S (2000) Pelicans and people: the two-tier village of Kokkare Bellur, Karnataka, India. Community based conservation in South Asia: case study no. 4. Kalpavriksh and International Institute of Environment and Development, New Delhi
Mundkur T, Carr P, Hean S, Somean C (1995) Survey for large waterbirds in Cambodia March–April 1994. IUCN, Gland
Pennant T (1790) Indian zoology. Henry Hughs for Robert Faulder, London
Price T (2008) Speciation in birds. Roberts & Company Publishers, Colorado
Pyhälä M (2001) Birds of Islamabad: status and seasonality. A WWF- Pakistan handbook. WWF, Lahore
Rahmani AR (1997) The effect of Indira Gandhi Nahar Project on the avifauna of the Thar Desert. J Bombay Nat Hist Soc 94:233–266
Rahmani AR (in press) Threatened birds of India and their conservation requirements. Bombay Natural History Society, Mumbai
Rasmussen PC, Anderton JC (2005) Birds of South Asia: the Ripley guide, vols 1 and 2. Lynx Edicions, Washington
Ripley SD, Beehler BM (1990) Patterns of speciation in Indian birds. J Biogeogr 17:639–648
Roberts TJ (1991) The birds of Pakistan. 2 vols. Oxford University Press, Karachi
Robson C (2000) A field guide to the birds of South-East Asia. New Holland, London
Salakij C, Salakij J, Narkkong N, Pitakkingthong D, Poothong S (2003) Hematology, morphology, cytochemistry and ultrastructure of blood cells in Painted Stork (*Mycteria leucocephala*). Kasetsart J (Nat Sci) 37:506–513
Santiapillai C, de Silva M, Dissanayake SRB, Alagoda TSB (2001) Aspects of the ecology of the Painted Stork (*Mycteria leucocephala*) in Ruhuna National Park. Tigerpaper 27:11–13
Scott DA, Rose PM (eds) (1989) Asian waterfowl census 1989. Mid-winter waterfowl counts in Southern and Eastern Asia January 1989. IWRB, Slimbridge
Sibley CG, Ahlquist JE (1990) Phylogeny and classification of birds: a study in molecular evolution. Yale University Press, New Haven
Sibley CG, Monroe BL Jr (1990) Distribution and taxonomy of birds of the world. Yale University Press, New Haven

Sibley CG, Ahlquist JE, Monroe BL Jr (1988) A classification of the living birds of the world based on DNA-DNA hybridization studies. Auk 105:409–423

Sivaperuman C, Baqri QH, Ramaswamy G, Naseema M (eds) (2009) Faunal ecology and conservation of the Great Indian Desert. Springer, Berlin

Slikas B (1997) Phylogeny of the avian family Ciconiidae (storks) based on cytochrome b sequences and DNA-DNA hybridization distances. Mol Phylogenet Evol 8:275–300

Slikas B (1998) Recognizing and testing homology of courtship displays in Storks (Aves: Ciconiiformes: Ciconiidae). Evolution 52:884–893

Smythies BE (1940) The birds of Burma. Oliver and Boyd, London

Subramanya S (1996) Distribution, status and conservation of Indian heronries. J Bombay Nat Hist Soc 93:459–480

Sundar KSG (2006) Flock size, density, and habitat selection of four large water birds species in an agricultural landscape in Uttar Pradesh, India: implications for management. Waterbirds 29:365–374

Urfi AJ (2010) Using heronry birds to monitor urbanization impacts: a case study of Painted Stork *Mycteria leucocephala* nesting in the Delhi Zoo, India. Ambio 39:190–193

Urfi AJ, Sen M, Kalam A, Meganathan T (2005) Counting birds in India: a review of methodologies and trends. Curr Sci 89:1997–2003

Wells DR (1999) The birds of the Thai-Malay Peninsula (covering Burma and Thailand south of the eleventh parallel, Peninsular Malaysia and Singapore) volume 1 (non-passerines). Academic, London

# Chapter 2
# Coloniality

> *...... the trees are literally covered with bird-life: every part of the crown is hidden by its noisy occupants who fight and struggle with each other for perches. Each tree appears like a moving mass of black, white and grey,...*[1]

**Abstract** The Painted Stork nests in mixed species colonies, called heronries, all across its range. Although avian coloniality is an active area of research, there is as yet no satisfactory explanation for the evolution of this trait. The first part of this chapter discusses the theoretical framework of avian coloniality while the second part is devoted to brief accounts of some of the well-known colonies of Painted Stork from India which are referred to from time to time in this book. For the purpose of convenience, these are grouped under three broad categories – heronries in urban parks and gardens, in marshes inside protected areas, and those in villages. The general patterns of population fluctuations and patterns of colonization are enumerated.

## 2.1 Avian Coloniality

After months of searing sun, when all is heat and dust, the arrival of the monsoon in India is an event. In June, dark, bloated clouds loom overhead, and when they finally release their load of rain, everyone welcomes the sweet musty smell of wet earth. The brown landscape suddenly turns green, and dried up rivers fill, swell, and overflow their banks. The rains trigger the spawning of fish, amphibians, and other aquatic creatures, which are carried far and wide by the flood waters. The surplus of edible freshwater species stimulates more new life in the wetlands, and fish-eating birds like the Painted Stork rush to take advantage of the bounty. During the summer, they hunt alone or in small feeding parties in ponds, marshes, and wet fields all across the countryside (Urfi 1998). But as the monsoon draws to a close, they start gathering at traditional nesting colonies. The question is why take the trouble of

[1] An Anonymous correspondent in Hume and Oates (1890).

A.J. Urfi, *The Painted Stork: Ecology and Conservation*,
DOI 10.1007/978-1-4419-8468-5_2, © Springer Science+Business Media, LLC 2011

flying large distances, possibly hundreds of kilometers, to reach certain sites, where others of their kind have also gathered? Surely, it should not be difficult to find a mate and also a suitable tree for building a nest in the vicinity of wherever they are. So the question really is why breed in colonies?

Coloniality is a perplexing phenomenon. Early British naturalists, who laid the foundations of ornithology in India, could barely conceal their bewilderment at the sight of heronries, as is evident from their writings. For instance, Betham (1904) wonders

> It is curious why waterbirds should breed in colonies; where trees are scarce, it might be understood, but where there are plenty standing in water, why they should pack together is hard to understand. It must be miserably uncomfortable to have no elbow room.

However, as a survival strategy, coloniality must have certain benefits for it to have evolved through natural selection. While there must be costs associated with nesting in the company of other birds too, the benefits must surely outweigh the costs.

Most species of birds are territorial and breed solitarily but a small number nest colonially. Lack (1968) estimated that approximately 13–14% of all birds are colonial, although this estimate is most probably conservative (Brown and Brown 2001). The tendency to breed in colonies seems to be more common in waterbirds. Among the Indian species of storks, Painted Stork, Asian Openbill, Lesser Adjutant, and Greater Adjutant are colonial nesters but the Woolly-necked Stork and the Black-necked Stork are not. So as a taxonomic group, not all storks are colonial and for finding answers to the evolution of coloniality we will have to look elsewhere, possibly in ecology and the way the resources that birds utilize are distributed in the environment (Urfi 2003b).

Much of our understanding about avian coloniality stems from the seminal work of the British Ornithologist David Lack (1966, 1968) who suggested that the major selective factors favoring colonial nesting in birds are chiefly enhancing avoidance of predators and enhancing more efficient exploitation of food resources. Since then, the subject of avian coloniality has been reviewed several times (Wittenberger and Hunt 1985; Siegel-Causey and Kharitonov 1990; Brown and Brown 2001; Burger 1981). Brown and Brown (2001) note that most research on avian coloniality has failed to uncover general rules that can easily explain either the evolution of coloniality or its maintenance. This is in contrast to say territoriality which can be explained in terms of economic defensibility and cooperative breeding, which can be understood in terms of "inclusive fitness theory."

### *2.1.1 Definitions*

The first question naturally is, how is coloniality defined? One definition, focusing on the clustering pattern of nests on trees is that a species is to be regarded as colonial if it clusters its nests enough to have measurable interactions with neighboring conspecifics while at the same time not defending feeding territories around the nest

(Brown and Brown 2001). Having said this, it should be mentioned that a fundamental question posed by ecologists (see Jovani and Tella 2007) is, how large a gap is necessary between nesting birds before one colony becomes two? (This point is revisited in Chap. 4.) Besides internest distances, other criteria, notably behavior, have also been used to designate colonies. So, another way to designate a group of nesting birds as a colony in the field is to see whether they respond simultaneously to the approach of a predator or depart from their nests simultaneously to forage (Brown and Brown 2001).

Coloniality can be "active" or "passive." If one looks around at a colony, then it will be immediately obvious that the canopies of some trees are densely packed with nests but in their immediate vicinity there are other trees, very similar to the ones bearing nests, which the birds have seemingly ignored (Betham 1904). Of course what appear to the human eye as "similar trees" may actually be very different substrates from the bird's view. (Therefore, the results of studies attempting to test differences in "selected" and "nonselected" trees on the basis of preselected criteria by the investigator have to be interpreted with care.) But it is clear that birds have in some sense chosen to nest in close proximity to each other and therefore, this is a case of "active coloniality," basically different from a situation where due to the scarcity of nesting sites, the birds have no option but to breed in colonies. The latter known as "passive coloniality," can happen in the case of seabirds, where suitable islands for building nests may be scarce, leading to dense concentrations of nests on islands. Such islands may be few and far between (Kharitonov and Siegel-Causey 1988). However, it must also be mentioned that the idea of the so-called "active coloniality" has been severely challenged. It has been argued that nesting aggregations may be merely a byproduct of many individuals selecting commodities such as habitat and mates, and not necessarily a benefit to the individuals that aggregate (Wagner et al. 2000).

In many parts of India, it is a common sight to see mynas and crows roosting together just as in many parts of Europe large roosts of starling are quite common. This is another type of group behavior with costs and benefits associated in a manner which is similar to group nesting. However, it is different from colonial nesting in a very fundamental way. Nesting colonies are what have been described as "central place systems," i.e., an individual belonging to the colony has to perforce return to it every day during the breeding season because it may have chicks or eggs in the nest. On the other hand, a roosting site is not a central place system in that an individual is not bound to return to it every day and can change its roost site every now and then. A whole body of theory has developed about central place systems, particularly in context of foraging ecology (Krebs and Davies 1984).

## 2.2 The Economics of Resource Defense

A good way to understand avian coloniality is through trying to understand territoriality and resource defense in birds. In order to understand the reasoning underlying territory sizes in birds and their defense (primarily food resources therein), simple

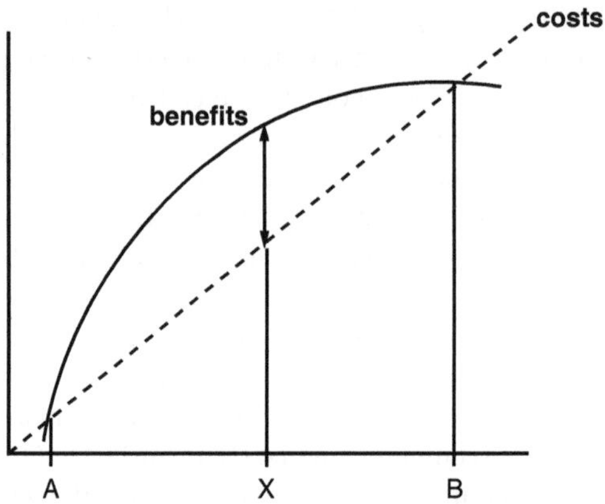


**Fig. 2.1** The economic defendability model. Horizontal axis shows the size of the territory and vertical axis is the amount of costs or benefits incurred/received. As the territory size increases both the costs and benefits increase but in different ways. With increasing territory size, benefits from territory acquisition (e.g., food) would be expected to increase initially but they will soon level off when the bird's needs have been met. On the other hand, the costs (e.g., energetic cost of defending the territory) would be expected to increase with increasing territory size but there would be no leveling off unlike the benefits. The territory should be defended between points A and B. The optimum territory size would be that at which the difference between the costs and benefits is greatest (point X in the figure). Redrawn from Perrins and Birkhead (1983)

graphical models based on optimality concepts are extremely useful (Urfi 2003b). Our starting premise is that resources are not evenly distributed in space and time. In fact, only three situations are possible, the first being that resources are randomly distributed, not a very realistic situation because resources like food, nesting material, or availability of mates are generally not randomly distributed. While resources are also not uniformly distributed, they are more likely to be distributed according to some pattern, which may be either spaced out or highly clumped (Krebs 1989). Imagine schools of fish in an ocean. There may be vast empty spaces where there will be no fish and then suddenly a school is encountered. From the view point of a predator, say, a sea-bird, vast distances have to be covered to track and locate these localized food resources but once sighted, there is no real scarcity because of the supra-abundance of fish locally.

Given these basic patterns of resource distribution, we can now ask the question: what sort of nesting patterns should evolve in different environments? In the case of regularly distributed resources, an individual may defend an all-purpose territory which has abundant food. If the resources are densely distributed, then territory sizes will be small and correspondingly, the density of breeding pairs high. However, if the resources are sparsely distributed, then territory sizes will be large and the density of nesting birds low. These concepts are illustrated in Fig. 2.1 where the costs of territory defense are shown to increase linearly with increasing territory size and the benefits shown to level off after a critical stage of territory size.

Imagine a flock of birds prospecting for a school of fish in the sea. Once the school is located, the birds get busy harvesting this resource. Since food is abundant, there is no point in defending this resource, which is also anyway impossible to defend by virtue of it being in the water. Also, since the food is located at a place where the nest cannot be built (all birds are terrestrial animals and require land for nesting purposes), the nesting and feeding areas are far apart. There would be a collapse of the territorial mode of nesting and nesting in colonies on the shore might be an option. Conspecifics flying in the direction of food resources, which are highly variable in space and time, may provide information, unintentionally and indirectly of course, to others about the location of food patches. If one bird is able to locate a food resource, others follow suit. In fact, some ornithologists have gone to the extent of suggesting that the plumage patterns of certain waterbirds may have evolved to facilitate increased communication regarding location of food sources. This may perhaps explain why many colonial species of herons, storks, etc. are uniformly colored in white or black.

There are other ways to understand how variations in the patterns of food distribution can influence the nesting behavior of birds. Perhaps, the most lucid of these is the geometric model (Wittenberger and Hunt 1985) which estimates the distance that will be traveled by a hypothetical bird to collect food in different situations. The model assumes two contrasting types of food distribution patterns, the stable and the variable food distribution patterns. The stable food situation could be a forest that has a number of trees, which a bird can exploit for food. Since all the requirements of a nesting pair can be met by, say, a certain minimum number of trees, it would be a sensible strategy to nest solitarily with a small all-purpose territory to defend, inside which only a certain distance has to be traveled for harvesting resources. As shown in Fig. 2.2, in this example, a total of four nests can constitute a territory. In terms of average foraging distance, this would be economically feasible compared to a situation in which all the four nests were located in the geometric center of the patch (the average distance to be traveled in the latter case being much higher, $D=2.94$ units of distance). However, the situation changes dramatically if the food distribution is variable. Seen purely in terms of the distance that needs to be covered for harvesting these resources, having all the nests placed in the geometric center of the colony is advantageous ($D=2.94$) as compared to having nests spaced out ($D=3.86$). In such a situation colonial nesting should evolve.

## 2.3 Evolution of Coloniality

The cost-benefit approach has dominated thinking on avian coloniality for a long time. A common approach has been to list out all the advantages and the disadvantages in colonial nesting. However, in this the problem is that many variables can enter the argument both as costs and benefits almost simultaneously (Wittenberger and Hunt 1985). For example, energetics can enter the argument as a benefit because colonial nesting is believed to enhance food finding and foraging efficiency.

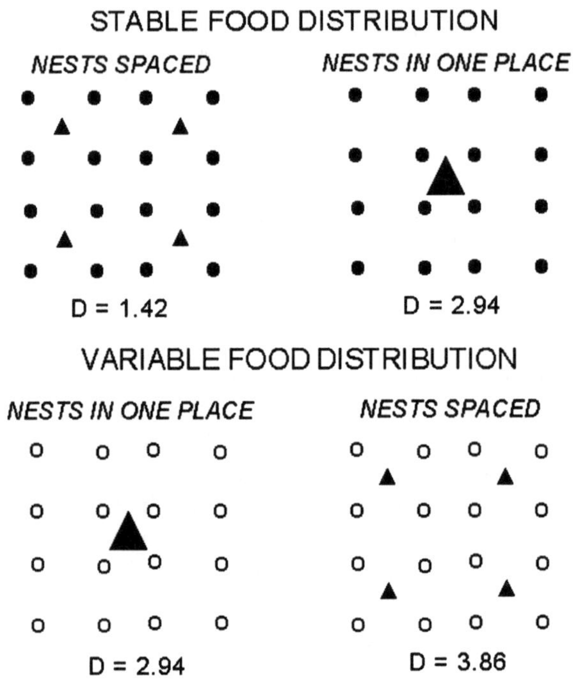


**Fig. 2.2** The geometrical model for avian coloniality. The *solid circles* represent resources that are continuously available. The *open circles* represent food resources that are variable in space and time, each of which is exploited to an equal extent during the course of the breeding season. *Small black triangles* represent single nests and the *larger black triangles* represent colonies with four nests. The model presents the mean travel distance for food (*D*) in different situations. *D* is lowest for dispersed nesters if the food distribution is stable, i.e., continuously available. However, when the food distribution is variable, then *D* is the lowest for colonial (central site) nesters. For details see text. Redrawn from Wittenburger and Hunt (1985)

However, the energetic aspects of food finding can also enter the argument as a cost because some amount of competition, including stealing of food, occurs both at the food patch as well as at the nesting site (also see Chap. 6). Another pressure on nesting birds is predation and this too can enter the argument both as a cost and a benefit. For instance, group living certainly increases group defense against predators. Hamilton (1971) has shown how the chances of an individual being predated upon decrease when it lives in flocks or groups. But group living has its disadvantages too, and one of them is increased visibility to predators. One of the unique aspects of colonial species is that colonies may be the only place where they can find mates because in the nonbreeding season, each individual bird, whether male or female, leads a solitary existence. At the colony site, there are a number of other benefits too (such as the ease with which nesting material can be stolen from neighboring nests) but the benefits come with a price tag attached.

Wittenburger and Hunt (1985) consider the net effect of coloniality on each significant variable. They argue that coloniality should evolve when

$$N_p + N_e + N_m + N_i > 0,$$

where $N_p$ is the benefit from enhanced defense minus the cost of increased attraction of predation (net effect on vulnerability to predation), $N_e$ is the benefit from enhanced foraging efficiency minus the cost of increased competition (net energetic effect), $N_m$ is the benefit of increased access to mates minus the cost of increased competition for mates (net effect on mating opportunities), and $N_i$ is the increased opportunities to exploit or disrupt neighbors (e.g., steal nest material or kill chicks) minus the increased degree of interference perpetrated by neighbors.

As a postscript it may be added that the cost-benefit approach outlined above has also been heavily criticized. For instance Danchin and Wagner (1997) have held that the orthodox economic framework has failed to provide a general explanation of colony evolution, chiefly because the balance of costs and benefits is extremely difficult to assess, certainly in the short term (Tella et al. 1998). A detailed discussion on the various theories of evolution of coloniality in birds is outside the scope of this book and we have to leave it at that, for the moment.

In India and other parts of Asia, colonies are generally mixed or multispecies colonies where several species of birds coexist in space and time, building nests on the same trees. These heronries are well structured with different birds occupying nesting places in relation to their body sizes, habits, etc. By employing optimality models, similar to those described above, Burger (1981) has attempted to demonstrate how mixed species colonies evolve.

## 2.4 Heronry, Rookery

Terms like heronry, rookery, pelicanry, etc. are often used while referring to colonies of waterbirds. It may be useful to clarify upon these terms and the connotations these words carry, before discussing other aspects of Painted Stork colonies.

While pelican colonies have been called "pelecanries" (Gee 1960), stork colonies probably never had a special name for themselves. A term such as "storkery" does not exist. Instead, the general term "heronry" has been used for referring to nesting colonies of storks, egrets, herons, cormorants, etc. It turns out that this word is a derivative of the word "rookery," i.e., a colony of rooks' nests. Till not so long ago heron colonies were called "heron rookeries." The term rookery has been widely used in biological literature since to describe colonies of animals as diverse as birds (crows, rooks, seabirds), marine mammals (sea lions), and even some turtles (Wikipedia 2010).

It turns out that the origins of the word rookery lie in Victorian age England. It was probably first used in print toward the end of the eighteenth century as a slangy term for a slum or what the dictionary describes as, "a cluster of mean tenements densely populated by people of the lowest class." In Victorian times slums or rookeries had become the most disreputable and notorious parts of many British cities. About the famous St. Giles Rookery (Fig. 2.3) in the heart of London, Charles Dickens wrote in "Sketches by Boz,"

**Fig. 2.3** An old sketch of the St. Giles rookery in London. *Source*: British History Online (2010)

> Wretched houses with broken windows patched with rags and paper: every room let out to a different family, and in many instances to two or even three ... filth everywhere — a gutter before the houses and a drain behind — clothes drying and slops emptying, from the windows; girls of fourteen or fifteen, with matted hair, walking about barefoot, and in white great-coats, almost their only covering; boys of all ages, in coats of all sizes and no coats at all; men and women, in every variety of scanty and dirty apparel, lounging, scolding, drinking, smoking, squabbling, fighting, and swearing.

The rook is a noisy and sociable bird and so it is hardly surprising that Victorian age writers and observers found the word for a colony of treetop nests, with loud, garrulous, squabbling owners described slums best. However, "heron rookeries" or simply heronries are not very pleasant places either and so they too lend themselves easily to negative images. This is especially so when there are chicks in the nests, constantly making irritating noises, begging their parents for food. Their droppings give the place a distinctly fishy odor, contributing to their repulsive character and there are reports from India of villagers cutting down trees on which waterbirds have formed nesting colonies (see Chap. 7 for more on this). Therefore, not surprisingly, heronries have been written about in somewhat disparaging terms by ornithologists. Jerdon (1864) referred to a pelican colony in South India as consisting of "rude nests." Ali and Ripley (1987), describing heronries at Keoladeo Ghana, wrote,

> ...often twenty nests or more on a single tree crowded cheek by jowl in disorderly tiers...

Sometimes, typically Indian terms have also been used in describing heronries. About White Ibis, which forms separate conclaves in mixed species heronries, Ali and Ripley (1987) noted that it has "a tendency to segregation [sic] into discrete *mohallas*." The context is right in that the ibis does segregate itself and forms subcolonies in a colony but the term used for the separate conclaves, "mohalla," is perhaps indicative of a stream of thought that subconsciously points toward preconceived notions and stereotypes. Mohallas refer to neighborhoods or localities in cities and towns of Central and South Asia. Often negative images have been associated with this term since it is often used derogatively to refer to conclaves of specific communities.

## 2.5 Some Painted Stork Colonies in India

There are huge variations in the sizes and types of Painted Stork colonies all across India. Single species (i.e., Painted Stork only) colonies are rare and mixed species heronries seem to be the norm. While most colonies are located on trees growing on islands, colonies on large trees in villages and agricultural fields are also known (Urfi 1993a). Colonies are found in the midst of dense human habitations as well as in pristine areas, inside protected areas, wildlife sanctuaries, and national parks. Most colonies are located on trees or at a height but nesting substrates vary considerably, from low level *Acacia* and *Prosopis* trees to large *Ficus* and *Tamarindus* trees. One of the lowest substrates recorded is *Euphorbia* stands at Man Marodi, an offshore island in the Gulf of Kutch (Urfi 2003a).

It is difficult to be sure exactly how many viable Painted Stork colonies exist. In literature one comes across very old heronries, with a history of existence of over 100 years or more as well as recently formed ones (Subramanya 1996).

As discussed in detail in the next chapter, there seems to be a definite concentration of Painted Stork colonies in western and northern India, and in South India. Certain states of India, notably Kerala on the extreme southwest and states of Bihar, Orissa, Jharkhand, West Bengal, Assam, and north eastern states of India, seem totally devoid of Painted Stork colonies. However, some of the same areas have colonies of Asian Openbill and Black-headed Ibis, which means that suitable nesting sites for large-sized heronry birds exist here. Probably, these areas lie outside the zone of permanent residence of Painted Stork (del Hoyo et al. 1992).

In storks food availability is the single most important factor governing all aspects of their ecology and nesting. In fact so crucial is the demand for suitable foraging grounds that some species are forced to commute large distances when food sources closer to the nesting colony have deteriorated and demand for food has reached a peak. Wood Storks have been recorded traveling to foraging sites about 130 km away from their colony (del Hoyo et al. 1992). Naturally, such long flights involve considerable expenditure of time and effort and so, if other conditions (viz. safety from ground predators, suitability of substrate, etc.) have been met then, colony sites close to the feeding grounds are preferred. In all the examples described below, the foraging grounds, either marshes associated with a river or irrigation tanks, are located within a reasonable distance from the colony site. However, the fact that the feeding areas lie away, though not too far away, from the nesting colony, has implications for conservation. This is so because if the conditions outside the premises are deteriorating, due to problems linked with pollution, encroachments upon natural habitats, urbanization, etc., then the birds are certainly at a risk. This, as we shall discuss in Chap. 8, is often the case.

In the remainder of this chapter, some well-known Painted Stork colonies in India, referred to from time to time in different contexts throughout this book, are described. These colonies are representative of a few broad categories under which most heronries in India seem to belong. The first category encompasses heronries in urban parks and gardens such as Delhi Zoo in New Delhi, Piele and other gardens in the city of Bhavnagar, and Karanj Tank in Mysore. The Negra Zoo in Kuala Lumpur, Malaysia, could also be included in this category. The second category includes heronries in protected areas such as Keoladeo Ghana National Park in Bharatpur, Rajasthan and Sultanpur National Park on the outskirts of Delhi. All of these heronries are located in the countryside and are therefore influenced by agricultural practices, land use patterns, and use of pesticides. The third category includes heronries that can best be described as "village heronries," i.e., colonies on trees in the middle of villages or near tanks and other waterbodies in the vicinity of villages, all across India. While there are several examples of such heronries from India, a description of the most famous one Kokkare Bellur is provided here. Another category of heronries includes those in village tanks.

### 2.5.1 Colonies in Urban Areas

*Delhi Zoo* (28°36′N 77°14′E) is spread over 85 ha and lies on the western bank of the river Yamuna in Delhi. It was created in 1959 as a national zoo (Central Zoo Authority 2009). Lying between two famous monuments, Old Fort (Fig. 2.4) and Humayun's Tomb, the zoo also has several historical structures scattered across its premises. However, from a biodiversity standpoint the unique feature of this zoo is the hundreds of wild waterbirds breeding in its four large ponds (Fig. 2.5), making it virtually a waterbird sanctuary.

**Fig. 2.4** The heronries of Delhi Zoo. Painted Stork nesting colonies are on clumps of mesquite trees growing on islands in the ponds of the zoo. The ramparts of the Old Fort can be clearly seen in the background. The picture was taken toward the commencement of the nesting season in September and the storks can be seen attending to nest building chores. Picture: A.J. Urfi (2010)

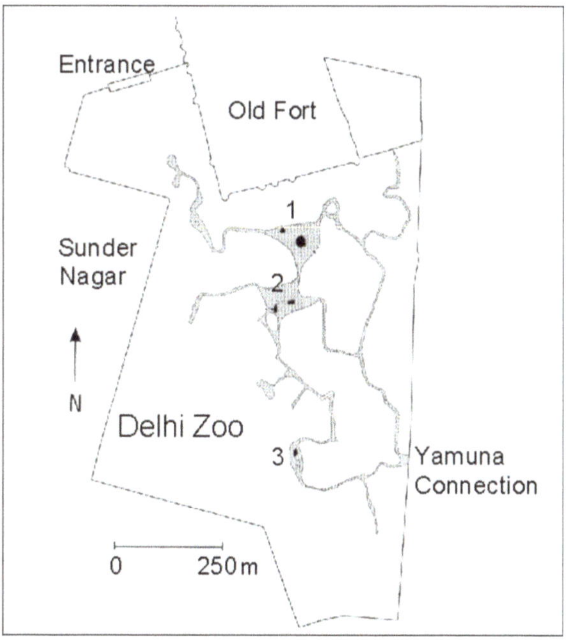


**Fig. 2.5** Map to show the layout of the Delhi zoo and its interconnected ponds where the nesting colonies of Painted Stork and other heronry birds are located. *Source*: Urfi (1997)

**Table 2.1** Different species of colonial waterbirds recorded nesting at some of the sites discussed in the text

| | Sites | | | | |
|---|---|---|---|---|---|
| Species | DZ[a] | Bhav[b] | SNP[c] | KDGNP[d] | KKB[e] |
| Darter | (–) | – | – | + | – |
| Little Cormorant | + | + | – | + | + |
| Indian Cormorant | + | – | + | + | – |
| Great Cormorant | (–) | – | – | + | – |
| Little Egret | + | – | – | + | + |
| Western Reef Heron | (–) | + | (–) | (–) | (–) |
| Great Egret | (–) | + | – | + | + |
| Intermediate Egret | + | – | – | + | + |
| Cattle Egret | + | + | + | + | + |
| Indian Pond Heron | + | + | (+) | (+) | + |
| Black-crowned Night Heron | + | + | + | + | + |
| Black-headed Ibis | + | + | + | + | + |
| Eurasian Spoonbill | (–) | + | – | + | – |
| Dalmatian Pelican | (–) | – | – | – | – |
| Spot-billed Pelican | (–) | – | – | – | + |
| Painted Stork | + | + | + | + | + |
| Asian Openbill | (–) | + | + | + | + |

+ recorded, (+) presumed to be present though not mentioned in the reference cited, (–) known to be absent from the site, – denotes that presence/absence data was not available in the source cited
[a]DZ (Delhi Zoo), Urfi (1997)
[b]Bhav (Gardens in Bhavnagar city), Parasharya and Naik (1990)
[c]SNP (Sultanpur National Park), Urfi et al. (2007)
[d]KDGNP (Keoladeo Ghana National Park), Ali and Vijayan (1983)
[e]KKB (Kokkare Bellur), Manu and Jolly (2000)

A wild population of Painted Stork established itself here in ca. 1960 and has been nesting more or less uninterruptedly since then (Urfi 1997). Every year after the closure of the Summer Monsoon, the storks start congregating in the zoo and by March or April they leave, dispersing in the countryside around Delhi. Besides the Painted Stork, several other species of colonial waterbirds (Table 2.1) also nest in the zoo ponds utilizing the same nesting substrate. While these birds are no doubt opportunistic nesters, their annual visitation to the zoo could be indicative of, besides an environment of safety prevailing within the confines of the zoo, a shortfall of natural nesting habitats in the countryside. The Painted Stork population of the Delhi Zoo has been studied sporadically, though in fair amount of detail, over the past 5 decades, chiefly during 1966–1971, 1988–1992, and 2004–2006. First, Desai and coworkers studied its general biology, covering nesting ecology (Desai et al. 1977), diet and foraging (Desai 1971; Desai et al. 1974), and growth and development of nestlings (Shah and Desai 1975a, b). Thereafter, more or less coinciding with the popular interest in India generated by the Asian Waterfowl Census in counting waterfowl (Urfi et al. 2005), studies on populations and colonization patterns of Painted Stork were undertaken (Urfi 1993b). Subsequently, investigations on foraging

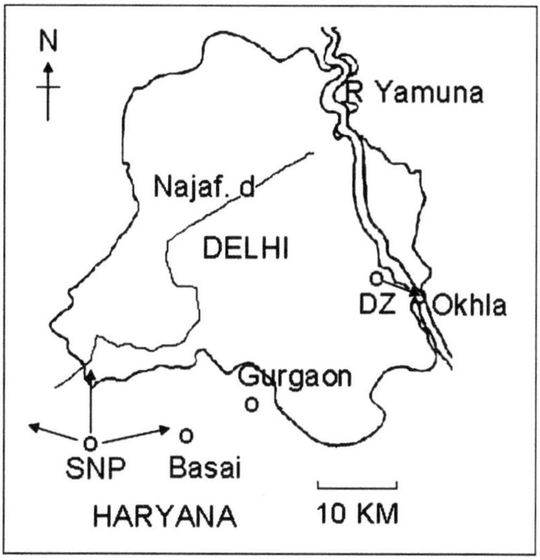


**Fig. 2.6** Map to show the location of Delhi Zoo and Sultanpur National Park and the known feeding grounds of the Painted Stork. Direction of foraging flights of Painted Stork to and from the nesting colonies is indicated by *arrows*. Note most of the foraging flights of the Delhi Zoo population are toward southeast in the direction of river Yamuna and the marshes associated with it, particularly Okhla Barrage Bird Sanctuary. At Sultanpur most flights are toward the Najafgadh drain, Basai, and an unidentified location toward the northwest. *Sources*: Urfi (1997) and Urfi et al. (2007)

ecology (Kalam and Urfi 2008), sexual size dimorphism and its consequences (Urfi and Kalam 2006), and intercolony variations in nesting parameters in the zoo colonies (Meganathan and Urfi 2009) were investigated. The Painted Storks of Delhi Zoo probably afford the best example of a detailed and consistent study on a single population of a wild bird from anywhere in India (Urfi 2010).

The zoo authorities put food for the birds (dead fish) in the ponds. Although the food is primarily intended for the pinioned exhibits in the zoo ponds, at the time when it is broadcast, all manner of fish-eating birds, including Painted Stork rush to grab this food. As far as those Painted Storks which are nesting in the zoo are concerned, there are reasons to believe that they hunt their own food and have little use for the dead fish put in the zoo ponds for feeding the exhibits. Soaring behavior (Desai 1971) indicative of flying out of the zoo to marshes associated with the river Yamuna for obtaining food of the right type and size is observed (Fig. 2.6). Those Painted Storks which are observed taking advantage of the free food offered by the zoo authorities are most probably nonbreeding individuals.

Since the Delhi Zoo colony has been in existence for so long, it would be expected that the behavior of birds has changed and they have adapted to the city environment. It has been pointed out that since the zoo Painted Stork population is based in a man-made setting and not in its natural habitat, many of the behavioral patterns would not be completely normal. This may be true to some extent because most birds exhibit a degree of behavioral plasticity due to which they are able to enhance their harvesting of resources. An interesting example in this regard is the use of anthropogenic thermals by Turkey Vultures, to extend their daily activity period, recorded at some landfill sites in Pennsylvania, USA (Mandel and Bildstein 2007). It may be mentioned here that sometimes, especially during cold winter days when the sun is not shining, the Painted Storks are observed circling in the vicinity of the

Indraprastha Thermal Power plant, located a few kilometers away from the zoo (A.J. Urfi unpublished information). Since this is a power plant, it is expected that the ambient temperatures may be higher than at other places and possibly the storks take advantage of the thermal column for gaining altitude. This is a matter of conjecture only, and it remains to be established if this really has become set as an adaptation in this population of Painted Stork.

*Bhavnagar city gardens*: Bhavnagar city (21′46′N, 72°11′E) is located in Gujarat state on the coast of the Gulf of Khambhat. The three distinct areas in the city where different species (Table 2.1) of colonial waterbirds nest are Old City area, Peile Gardens and its immediate vicinity, and the suburban areas of Krishnanagar and Takhteshwar (Parasharya and Naik 1990). About 16 different species of trees have been recorded to be used by Painted Stork for nesting. The common ones are Neem, Peepul, Tamarind, *Peltophorum pterocarpus*, *Skerculia foetida*, etc. The birds make frequent foraging trips to foraging areas, both freshwater and coastal sites, lying at varying distances from their breeding colonies in the city gardens.

*Karanji tank*: This tank in Mysore city (12°18′N, 76°42′E) of Karnataka has a spread of about 100 acres and is located on the eastern edge of the Mysore Zoo, formally known as "Chamarajendra Zoological Gardens." It has about nine man-made islands on which *Acacia arabica* and *Ficus* trees have been planted. Some of these trees, as well as those on the shore, are utilized by Painted Stork, Spot-billed Pelican, and other species of colonial waterbirds for nesting (Jamgaonkar et al. 1994). Birds have been observed flying in and out of the tank to areas where food is found.

### 2.5.2 Colonies in Marshes/Protected areas

*Sultanpur National Park*: Sultanpur (28°28′N 76°53′E) is located ca. 25 km to the southeast of Delhi in Gurgaon district of Haryana in a predominantly agricultural landscape, crisscrossed by irrigation canals. The area was notified as a bird sanctuary by the Haryana State government in 1971. Later, an area of 13,727 ha, including as its core ca. 143 ha of low-lying marshes, was declared a National Park (Islam and Rahmani 2004).

The dominant terrestrial vegetation in this area consists of trees of mesquite, Tamarind, Neem, *Acacia nilotica*, and several types of grasses. The plants recorded in the marshy areas of the park include submerged vegetation such as *Vallisneria natans* and *Ceratophyllum demersum*, emergents such as Elephant Grass, *Saccharum munja*, and *Cyperus rotundus*, and surface vegetation such as *Nymphaea stellata*, *Nymphaea nouchali*, and *Ipomoea reptans*.

The wetland has undergone several changes since the early 1970s (Ganguli 1975) when it became a popular bird watching site, frequented by birdwatchers from Delhi. Then it was a shallow depression, consisting of brackish water and attracted a large number of Lesser Flamingo. However, as a part of its management policy, the state forest department embarked upon a tree plantation program during the 1980s.

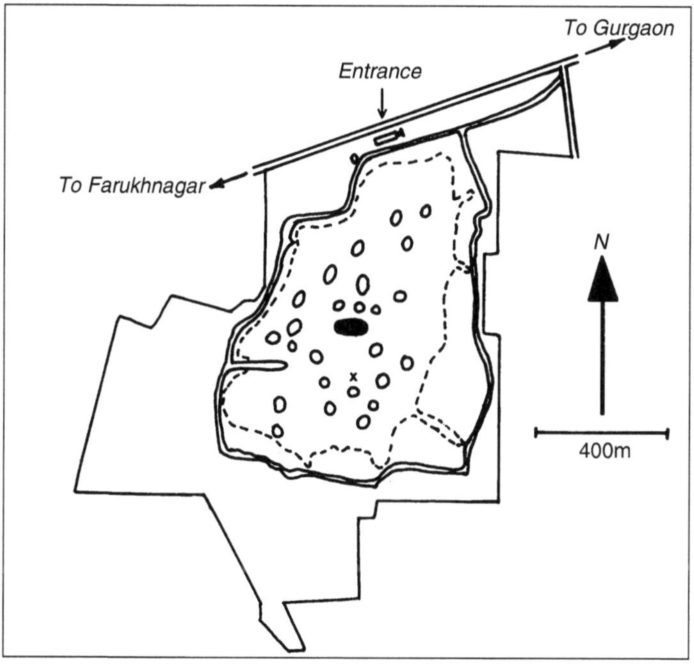


**Fig. 2.7** Sketch map of Sultanpur Bird Sanctuary to show the location of heronries in the marsh. The main heronry is shown in *black*. The area enclosed by the line - - - - encompasses the marsh. == around the marsh denotes the footpath which encircles the marsh. The line _____ denotes the boundary of the park. Note: Not all the islands/mounds have been shown. Point X denotes the point from which observations made. *Source*: Urfi et al. (2007)

Several mounds were created in the lake and *A. nilotica* trees planted on them. There are now about 50 islands, including one large island (ca. 100×40 m) in the center of the marsh (Fig. 2.7). Painted Storks nest on this island and on some smaller islands with canopy diameter <10 m, along with other species of colonial waterbirds (Table 2.1).

Painted Storks were first observed breeding in Sultanpur in 1993, when 40 nesting pairs were recorded (Poole 1994). Thereafter, their nesting here was sporadic, until around 2000, after which it occurred annually. Although at Sultanpur the heronry has got well established now, changes in the immediate environment of the park such as loss of wetland habitats, increase of built-up areas, noise, traffic, and pollution are a matter of concern (Urfi et al. 2007). The original *jheel* system of which Sultanpur was a part seems to have been greatly affected by these developments (Gaston 1994).

Field studies have indicated that Painted Stork nesting in Sultanpur head out in the direction where the marshes alongside the Najafgarh drain are located (Fig. 2.6). They are also observed flying in the direction of Basai wetland (Islam and Rahmani 2004) and an unknown destination lying towards the west. All of these outlying areas, beyond the park premises, are undergoing transformation due to urbanization and loss of wetlands.

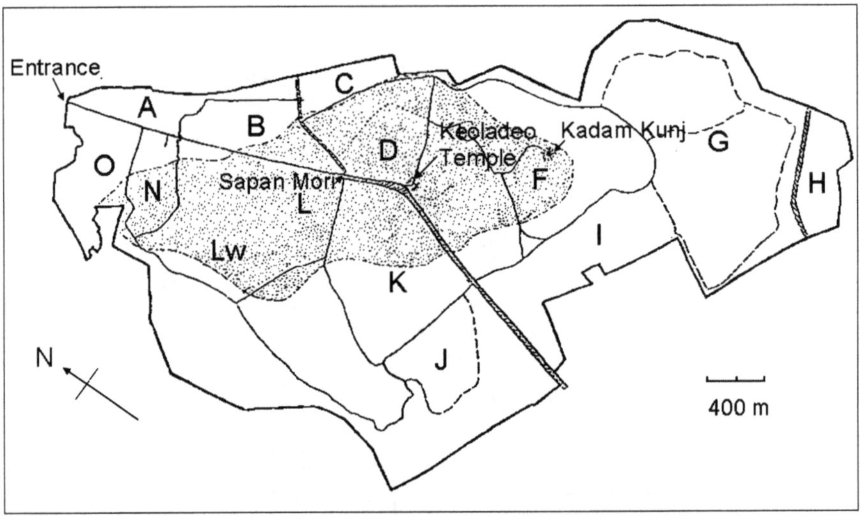


**Fig. 2.8** Sketch map to show the layout of Keoladeo Ghana National Park. Note that the whole marsh is bisected into a number of blocks by bund walls. The blocks which are favorites of Painted Stork are B, L, D. Each of the blocks has *Acacia* trees growing on mounds in the marsh which are utilized for nesting by colonial waterbirds. *Source*: Ali and Vijayan (1983)

*Keoladeo Ghana National Park*: This (27°7.6′ [to 12.2] N, 77°29.5′ [to 33.9′] E) 29 $km^2$ park is a UNESCO world heritage site as well as a Ramsar site. It used to be the only wintering ground for the western race of the endangered Siberian Crane, till some decades ago. Now of course this crane has stopped coming to India altogether. However, Keoladeo is still an important refuge for migratory waterfowl and is also famous for its extensive heronries (Table 2.1).

Keoladeo was once the private hunting ground of the local ruler, the Maharaja of Bharatpur. Large stone slabs at the Keoladevi temple, from which the park derives its name, with figures of bags etched on them, stand as a testimony to earlier times when shooting and hunting was in vogue. A system of dykes and bunds was created to store the surplus waters of the river Gambhir, and rows of *Acacia* trees were planted on the bunds and also on raised mounds in the middle of the marshes. As Fig. 2.8 shows, the park is bisected into a number of blocks by a series of bunds. Heronry birds build their nests on trees planted on the bunds and on islands in the marshes (Sankhala 1990). While in the past Painted Storks have built nests in several blocks, particularly L, B, and D, in the past few years nesting seems to be restricted only to blocks B and D.

Keoladeo was the site for a number of significant ornithological projects undertaken by the BNHS under the stewardship of the charismatic ornithologist Dr. Salim Ali. These projects include studies on ringing and migration of birds and also studies on hydrobiology of the Keoladeo wetland ecosystem (Ali and Vijayan 1983).

**Fig. 2.9** The signboard pointing toward Kokkare Bellur Pelicanry on the Mysore Bengaluru highway. Picture: A.J. Urfi (2001)

### 2.5.3 *Village Heronries*

*Kokkare Bellur*: Popularly known as the "stork village," Kokkare Bellur village (12°13′N, 77°05′E) is situated about 80 km from the city of Bangalore, in Mandya district of the south Indian state of Karnataka (Fig. 2.9). The area is a typical dryland village in southern India with its cultivated and fallow fields, cactus hedges, and trees in the fields and villages. The details of the flora of the site are given in Manu and Jolly (2000). Kokkare Bellur is bounded on the south side by the Shimsha River although it is some distance away. Within a 100 km radius of the village lie numerous irrigation tanks – principally Sule Kere, Malavalli Kere, Koppe Kere, Marehalli Kere, Shetty Kere, and Karanji Kere which are important feeding grounds for the birds nesting at Kokkare Bellur.

The checklist of the birds of Kokkare Bellur lists 141 species. A large number of trees line the huts and habitation of the villagers which are used by several species of colonial waterbirds for nesting (Table 2.1). The two main species nesting at Kokkare Bellur are Painted Stork and Spot-billed Pelican and during the period they are in residence in the village, i.e., from December to June, it appears as if the village is two-tiered (Fig. 2.10). On the ground are the humans and up in the trees there is a village of nesting birds, where the feathered bipeds carry on with their normal

**Fig. 2.10** A picture of Kokkare Bellur village showing the village houses and the trees used by the nesting storks and pelicans. Picture: A.J. Urfi (2001)

activities of breeding, feeding, etc., seemingly oblivious to the village below. It has been recorded that "even in the midst of wedding celebrations, when loudspeakers blast music all over the village, the raucous sounds of the birds, the quiver of heavy wings and the clatter of stork bills continue unaffected."

It is assumed that storks and pelican have been nesting here for a long time though since precisely when is not clear. Village legends put it to hundreds of years. Interestingly, the name of the village *kokkare*, which is the local name for stork, bears this out. The Kokkare Bellur colony came to the notice of local birdwatchers in the 1970s (Neginhal 1977) and subsequently its importance was highlighted (Nagulu and Rao 1983). It is widely held that in 1864 when T.C. Jerdon made the following observation (below) about a heronry in South India, but which he did not name, he probably meant it to be the village of Kokkare Bellur,

> I have visited one Pelicanry in the Carnatic, where the pelicans have (for ages I was told) built their rude nests, on rather low trees in the midst of a village, and seemed to care little for the close and constant proximity of human beings.

**Fig. 2.11** The heronry at Traj is a typical irrigation tank type with an island planted with trees. Picture: A.J. Urfi (1999)

### *2.5.4 Other Types of Colonies*

Some of the commonest types of heronries encountered in India today are those located in the middle of irrigation tanks. Seelaj (Urfi 2003c), Traj (Urfi 2002), Bhadalwadi Tank (Pande 2006), etc. are good examples of such heronries. Typically, a large irrigation tank gets filled up with water during the monsoon. In the middle of the tank there is an island or several islands, on which clusters of trees (either *Acacia* or *Prosopis*) are utilized as a substrate for nesting by Painted Stork and other birds (Fig. 2.11). In some cases these tanks are also areas of high biodiversity value. For instance, Traj irrigation tank in Gujarat has a small population of Mugger Crocodile, besides the heronry of course. Many irrigation tanks in cities probably hark back to the days when the city had not grown so much, and the tank used to be an integral part of the village.

Another type of colony is that on islands in the sea, as in the case of Man-Marodi island in the Gulf of Kutch, off the coast of Gujarat (Urfi 2003a). Mangrove colonies of Painted Stork are probably few.

## 2.6 Typical Patterns of Heronries

The populations of nesting birds at colonies are dynamic, and huge fluctuations in numbers are recorded diurnally as well as seasonally. At the Delhi Zoo there is a marked increase in numbers of Painted Stork in the months August–October when

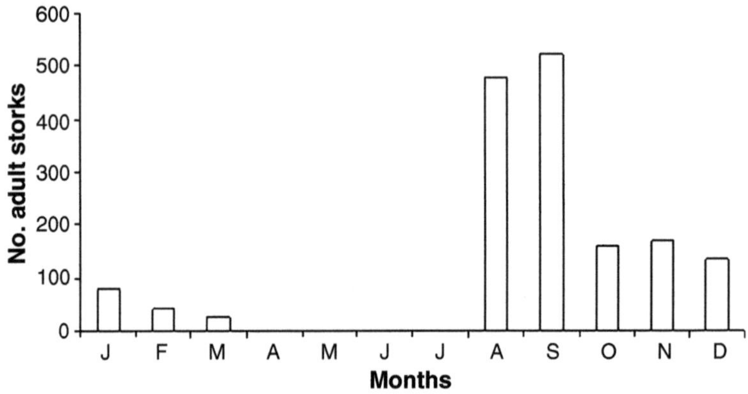


**Fig. 2.12** Roost counts of adult Painted Stork in different months of the year in the Delhi Zoo. Note, in certain months of the year storks are absent from the zoo. Immigration starts in August and peak numbers are reached in September. *Source*: Urfi (1997)

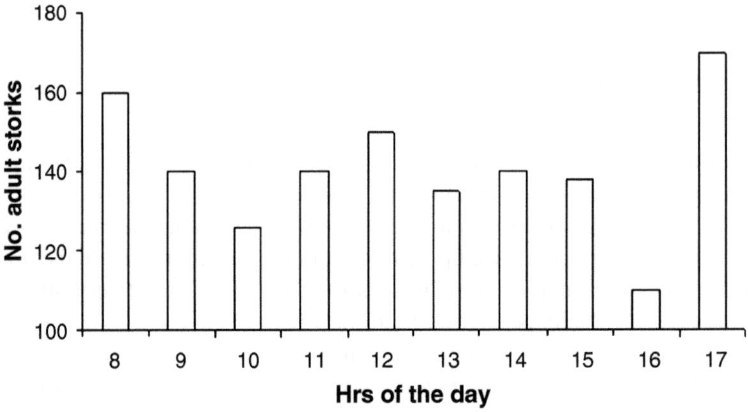


**Fig. 2.13** Roost counts of Painted Stork at different times of the day at the Delhi Zoo. Note considerable fluctuations in the number of birds at different times of the day. *Source*: Urfi et al. (2005)

the birds have congregated for nesting (Fig. 2.12). Although many species of colonial waterbirds are both colonial nesters and roosters (Gadgil and Ali 1975), the Painted Storks leave the site once the nesting is over and disperse in the countryside. Hence, in the summer the Delhi Zoo bears a deserted look.

During the course of a day, the numbers of birds counted are not constant (Fig. 2.13). Throughout the day the birds keep flying in and out of the premises on food finding missions or searching for nesting material. Typically, there is a period in the forenoon when most birds are away from the colony and another period in the afternoon, as well, when their roost counts are considerably low. From a census point of view, the best time to count birds is late evening when most birds return and stay at the nest (Urfi et al. 2005).

## 2.7 Colonization Patterns over the Long Term

At any site, colonization patterns vary considerably from year to year. For instance at the Delhi Zoo, the two islands in pond 1 were initially colonized by the founders in 1960 and probably for several years thereafter these colonies were the only colonies in the zoo (J.H. Desai, personal communication). In recent times, in years of normal rainfall the two islands in each of the two ponds (1 and 2) are colonized (Fig. 2.14a). Our observations are that pond 2 is colonized first (Meganathan and Urfi 2009) which is probably a reversal from the earlier trend during the 1960s when it seems pond 1 was the main heronry and was always colonized first. In years of exceptional rainfall, as happened in 1988 (Urfi 1997), Painted Storks built nests not just on the four main islands in ponds 1 and 2 but also on a number of other trees on the shore. Interestingly, pond 3 was also colonized during this year, something which does not happen in normal years (Fig. 2.14b).

From extensive surveys of Wood Stork colonies in southeastern United States, Frederick and Meyer (2008) concluded that colonies were not permanent. According to their study, older and larger colonies had survival rates compared to younger, smaller ones. Also, there appeared to be a direct relationship between the size of a

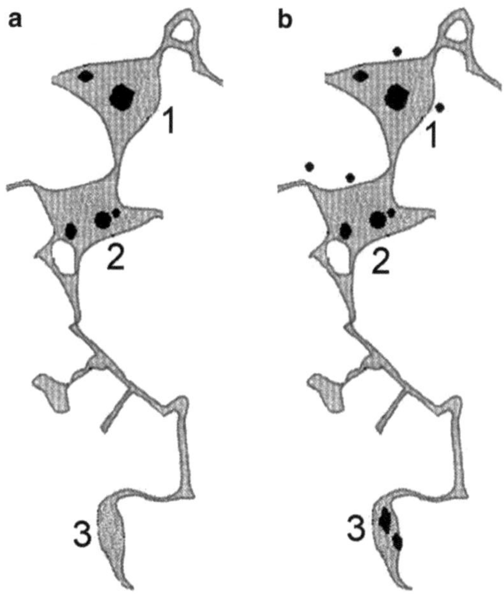


**Fig. 2.14** Figure to show the three interconnected ponds (1, 2, 3) of the Delhi Zoo. The islands where Painted Storks build nests are shown in *black*. (**a**) Shows the situation in years of normal rainfall and (**b**) shows the situation in years of above average rainfall in which large numbers of Painted Storks come to the zoo for nesting. Note in such years pond 3 is also colonized which is not the case in normal years

**Fig. 2.15** The status of some Painted Stork colonies in the Delhi region, along the river Yamuna. Symbols used *filled squares* extinct; *filled circles* extant, and *filled triangles* new. . — — — state boundary. Map not drawn to absolute scale

colony and longevity, hinting that larger colonies were also more stable breeding sites. As far as the Painted Stork colonies are concerned we do not have a clear picture yet. For the Delhi region, the picture which seems to emerge from the limited information available is described below.

In this region (Fig. 2.15) extinct, extant, and fresh Painted Stork colonies all seem to be located along the river Yamuna basin. Among the oldest and now extinct colony is the one reported at Goverdhan, close to the city of Mathura. According to Hume and Oates (1890), this colony was located on four trees, three of Tamarind and one of *Ficus* sp. and had about 70 nests of Painted Stork. MacDonald (1962) mentions about a Painted Stork colony, presumably on a large tree somewhere in or around Faridabad. Since there has been no news of this colony in recent years, it is presumed to be extinct. Another colony consisting of about 40 nests of Painted Stork on a large *Ficus* tree in the middle of an agricultural field was recorded near Bijana village near Sonipat, close to the western Yamuna canal (Urfi 1993a). Since there is no news of this colony, which was a colony of about 40 Painted Stork nests along with nests of some other birds on a huge Ficus tree in the middle of an agricultural field, it is presumed to be extinct. Interestingly, at the time when the above report was made, it was learnt from local villagers that the colony was sporadic and the birds did not build nests here every year.

While the Delhi Zoo has remained stable for the past 5 decades, since 1960 when it was formed, two new colonies have come up in the region. Sultanpur, by all accounts (Ganguli 1975), never had a Painted Stork colony and the present one came up only in 2000. Another new colony in this region is the one that has come up in the premises of the Oil refinery at Mathura (A.R. Rahmani, personal communication), which is probably a typical urban heronry in that birds like Painted Stork and other species breed on trees planted on islands, much like they do in Delhi Zoo.

## References

Ali S, Ripley SD (1987) Compact handbook of the birds of India and Pakistan. Oxford University Press, Delhi

Ali S, Vijayan VS (1983) Hydrobiological (ecological) research at Keoladeo National Park, Bharatpur (first interim report). Bombay Natural History Society, Mumbai

Betham RM (1904) Notes on bird's nesting from Poona. J Bombay Nat Hist Soc 15:709–712

British History Online (2010) The "Rookery," St. Giles, 1850 from: "Bloomsbury," old and new London: volume 4 (1878), pp 480–489. URL: http://www.british-history.ac.uk/report.aspx?compid=45209. Accessed 13 Nov 2010

Brown CR, Brown MB (2001) Avian coloniality, progress and problems. In: Nolan V Jr et al (eds) Current ornithology. Kluwer, New York

Burger J (1981) A model for the evolution of mixed-species colonies of ciconiiformes. Quart Rev Biol 56:143–167

Central Zoo Authority (CZA) (2009) Central zoo authority. Statutory body under the Ministry of Environment and Forests, Government of India. Central Zoo Authority, New Delhi. http://www.cza.nic.in. Accessed Aug 2009

Danchin E, Wagner RH (1997) The evolution of coloniality: the emergence of new perspectives. Trends Ecol Evol 12:342–347

del Hoyo J, Elliott A, Sargatal J (eds) (1992) Handbook of the birds of the world, vol 1. Lynx Edicions, Barcelona

Desai JH (1971) Feeding ecology and nesting of painted stork Ibis leucocephalus at the Delhi zoo. Int Zoo Yearbook 11:208–215

Desai JH, Shah RV, Menon GK (1974) Diet and food requirement of painted stork at the breeding colony in the Delhi zoological park. Pavo 12:13–23

Desai JH, Menon GK, Shah RV (1977) Studies on the reproductive pattern of the Painted Stork, *Ibis leucocephalus*, Pennant. Pavo 15:1–32

Frederick P, Meyer KD (2008) Longevity and size of wood stork (*Mycteria americana*) colonies in Florida as guides for an effective monitoring strategy in the southeastern United States. Waterbirds 31:12–18

Gadgil M, Ali S (1975) Communal roosting habits of Indian birds. J Bombay Nat Hist Soc 72:716–727

Ganguli U (1975) A guide to the birds of the Delhi area. Indian Council of Agricultural Research, New Delhi

Gaston AJ (1994) Some comments on the "revival" of Sultanpur lake. Orient Bird Club Bull 20:49–50

Gee EP (1960) The breeding of the Grey or Spot billed Pelican, Pelecanus philippensis Gmelin. J Bombay Nat Hist Soc 57:245–251

Hamilton WD (1971) Geometry for the selfish herd. J Theor Biol 31:12–45

Hume AO, Oates EW (1890) The nests and eggs of Indian birds, vol III. Taylor & Francis, London

Islam MZ, Rahmani AR (2004) Important bird areas in India: priority sites for conservation. Bombay Natural History Society, Mumbai

Jamgaonkar AV, Jacob PG, Rajagopal SR, Bhat HR (1994) Records of new breeding colonies of Painted Stork *Mycteria leucocephala* in Karnataka. Pavo 32:59–62

Jerdon TC (1864) The birds of India, vol III. George Wyman & Co., Calcutta

Jovani R, Tella JL (2007) Fractal bird nest distribution produces scale-free colony sizes. Proc Biol Sci 274:2465–2469

Kalam A, Urfi AJ (2008) Foraging behavior and prey size of the painted stork (*Mycteria leucocephala*). J Zool 274:198–204

Kharitonov SP, Siegel-Causey D (1988) Colony formation in seabirds. Curr Ornithol 5:223–272

Krebs CJ (1989) Ecological methodology. Harper and Row, Philadelphia

Krebs JR, Davies NB (eds) (1984) Behavioral ecology: an evolutionary approach. Blackwell Scientific Publications, London

Lack D (1966) Population studies of birds. Clarendon, London

Lack D (1968) Ecological adaptations for breeding in birds. Chapman and Hall, London

MacDonald M (1962) Birds in the sun: some beautiful birds of India. D.B. Taraporevala Sons, London

Mandel JT, Bildstein KL (2007) Turkey vultures use anthropogenic thermals to extend their daily feeding activity period. Wilson J Ornithol 119:102–105

Manu K, Jolly S (2000) Pelicans and people: the two-tier village of Kokkare Bellur, Karnataka, India, community based conservation in South Asia: case study no. 4. Kalpavriksh & International Institute of Environment and development, New Delhi

Meganathan T, Urfi AJ (2009) Inter-colony variations in nesting ecology of Painted Stork (*Mycteria leucocephala*) in the Delhi Zoo (North India). Waterbirds 32:352–356

Nagulu V, Rao JVR (1983) Survey of South Indian pelicanries. J Bombay Nat Hist Soc 80:141–143

Neginhal SG (1977) Discovery of a pelicanry in Karnataka. J Bombay Nat Hist Soc 74:169–170

Pande S (2006) Bhadalwadi Tank, a refuge for Painted Storks. Hornbill April–June:11–15

Parasharya BM, Naik RM (1990) Ciconiiform birds breeding in Bhavnagar city, Gujarat. In: Daniel JC, Serrao JC (eds) Conservation in developing countries: problems and prospects. Bombay Natural History Society, Mumbai

Perrins CM, Birkhead T (1983) Avian ecology. Blackie, London

Poole C (1994) Sultanpur lake revived. Orient Bird Club Bull 19:15

Sankhala K (1990) Gardens of God. The waterbird sanctuary at Bharatpur. Vikas Publishing House, New Delhi

Shah RV, Desai JH (1975a) Growth and development of the Painted Stork, *Ibis leucocephala*, Pennant. I. Embryonic development. Pavo 13:83–87

Shah RV, Desai JH (1975b) Growth and development of the Painted Stork, *Ibis leucocephala*, Pennant. II. Post hatching growth pattern and motor development. Pavo 13:88–101

Siegel-Causey D, Kharitonov SP (1990) The evolution of coloniality. Curr Ornithol 7:285–330

Subramanya S (1996) Distribution, status and conservation of Indian heronries. J Bombay Nat Hist Soc 93:459–480

Tella JL, Hiraldo F, Donázar JA (1998) The evolution of coloniality: does commodity selection explain it all? Trends Ecol Evol 13:75–76

Urfi AJ (1993a) Heronries in the Delhi region of India. Orient Bird Club Bull 17:19–21

Urfi AJ (1993b) Breeding patterns of Painted Stork (*Mycteria leucocephala*) at Delhi Zoo, India. Colonial Waterbirds 16:95–97

Urfi AJ (1997) The significance of Delhi Zoo for wild waterbirds, with special reference to the Painted Stork *Mycteria leucocephala*. Forktail 12:87–97

Urfi AJ (1998) A monsoon delivers storks. Nat History 107:32–39

Urfi AJ (2002) A heronry at Traj in Kheda district, Gujarat. J Bombay Nat Hist Soc 99:113–114

Urfi AJ (2003a) Record of a nesting colony of Painted Stork *Mycteria leuccephala* at Man-Marodi Island in the Gulf of Kutch. J Bombay Nat Hist Soc 100:109–110

Urfi AJ (2003b) Breeding ecology of birds. Why do some species nest singly while others are colonial? Resonance 8:22–32

Urfi AJ (2003c) The ornithological and conservation significance of village pond heronries. A case study of Seelaj and Traj village ponds around Ahmedabad (Gujarat). Mayura 16:3–9

Urfi AJ (2010) Using heronry birds to monitor urbanization impacts: a case study of Painted Stork *Mycteria leucocephala* nesting in the Delhi Zoo, India. Ambio 39:190–193

Urfi AJ, Kalam A (2006) Sexual size dimorphism and mating pattern in the Painted Stork (*Mycteria leucocephala*). Waterbirds 29:489–496

Urfi AJ, Sen M, Kalam A, Meganathan T (2005) Counting birds in India: a review of methodologies and trends. Curr Sci 89:1997–2003

Urfi AJ, Meganathan T, Kalam A (2007) Nesting ecology of Painted Stork (*Mycteria leucocephala*) at Sultanpur National Park (Haryana), India. Forktail 23:150–153

Wagner RH, Danchin E, Boulinier T, Helfenstein F (2000) Colonies as by products of commodity selection. Behav Ecol 11:572–573

Wikipedia (2010) Rookery. http://en.wikipedia.org/wiki/Rookery. Accessed 6 Oct 2010

Wittenberger JF, Hunt GL Jr (1985) The adaptive significance of coloniality in birds. In: Farner DS, King JR (eds) Avian biology, vol 8. Academic, San Diego

# Chapter 3
# Nesting Ecology

*Although many legends about the white stork's ability to supply human babies have been passed on for generations, sustained as cute and convenient ways of avoiding early sex education, relatively little has been written about how storks themselves propagate.*

Philip Kahl (1971a)

**Abstract** The nesting ecology of birds is understood in relation to proximate and ultimate factors. This chapter examines the nesting time of the Painted Stork across India and explores the significance of the monsoon in triggering off its nesting cycle. Also examined in this chapter are courtship patterns of the Painted Stork, nest building, nesting behavior, and particularly carrying fresh leafy twigs for nest lining. The chapter ends with an examination of habitat selection patterns, nest mortalities, and the role of predation and other agents.

## 3.1 Proximate and Ultimate Factors

Breeding is the most crucial phase in the life cycle of a living organism. In birds, generally speaking, as soon as a male has reached maturity and is ready to breed he tries to first establish a territory and having once secured it, starts looking for a mate. To use a human analogy, an individual in possession of a territory is, in a sense, a man of means. Nowhere has this phase, when the male is standing on the altar of marriage, been expressed more succinctly than in the opening lines of Jane Austen's classic novel "Pride and Prejudice,"

> It is a truth universally acknowledged, that a single man in possession of a good fortune, must be in want of a wife.

Beneath the veneer of the somewhat comic patterns of courtship behavior, more fundamental though extremely complex processes are at work. So, by the time an individual has come to a stage when it is on the look out for a mate, several physiological

A.J. Urfi, *The Painted Stork: Ecology and Conservation*,

DOI 10.1007/978-1-4419-8468-5_3, 

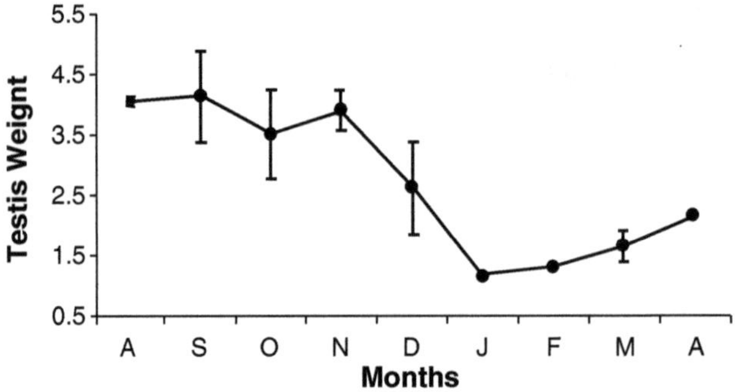


**Fig. 3.1** Monthly changes in the weight of testis (g ± sd) of Painted Stork at the Delhi Zoo during the breeding season 1970–1971. Redrawn from Desai et al. (1977)

processes inside the body of the bird, preceding the stage of mating and reproduction, have already been initiated. These are linked to maturation and growth of germ cells, mobilization of resources from body tissues toward gonad build up, formation of nutrient reserves such as yolk in the egg, for the embryo to grow and survive, etc. For a long time ornithologists have been trying to understand these processes, and address the all important question, what factors trigger off these events?

Since both birds and mammals are warm-blooded vertebrates, it is not surprising that comparisons between the two, with reference to their reproductive physiology, are often made. Perhaps the most obvious difference between them is the way in which gonads mature and develop. In mammals the gonads are more or less fully developed throughout the year, but in birds they enlarge only at the time of reproduction, only to regress soon after the reproductive phase is over. In this the Painted Stork is no exception since the same basic pattern of seasonal fluctuations in gonadal mass has been demonstrated in this species (Desai et al. 1977). As Fig. 3.1 shows, the testis are already enlarged in the month of August, when Painted Storks start arriving at the Delhi Zoo. From November onward, the mean testis weight decreases sharply, indicating regression. These changes are obviously a reflection of the fact that birds fly and the energetic demands of flight require that body mass be at a premium. Hence gonads are regressed outside the breeding season as much as possible. Interestingly, bats also show a greater degree of gonadal regression compared to other mammals (Dawson 2002).

From a physiological point of view, three phases in the annual calendar of birds are noteworthy. They are breeding, molting, and migrating. Since all of these are energetically demanding processes and temporarily reduce the activity of birds, they cannot obviously occur simultaneously. Therefore, in most species, one phase succeeds the other and there is no overlap. Generally, these phases are timed with seasons, upon which depend other crucial factors.

The Painted Stork, unlike say the White Stork, is not a migratory species, although it probably wanders across vast areas and covers huge distances in the process (Chap. 1). So essentially no physiological preparedness for migration should be expected in this case. As far as molting is concerned, overall, this aspect has been relatively poorly studied in storks (del Hoyo et al. 1992) and so very little is known about the Painted Stork. However, most probably there is no definite period when storks shed off their feathers. In fact, the more likely scenario is that they do not shed all their feathers completely in the manner of some waterfowl. So as far as resident birds like the Painted Stork are concerned, its annual calendar has only two phases, the breeding phase and nonbreeding phase.

Why does a particular species breed only at a certain time of the year? Why not at some other time? Is breeding restricted to a narrow time slot in a year or is it spread out across the season? Understandably, considerable interest has focused on the timing of breeding in birds and two terms, proximate and ultimate factors, have been frequently used in this context. The former includes a host of environmental factors that trigger off the internal processes related to gonad maturation and the initiation of mating behavior patterns. In many temperate zone countries, the environmental triggers that kick start the physiological preparedness for breeding and migration have been well studied. For instance, the amount of sunlight or changes in day length are known to be significant (Dawson 2002). In tropical countries, the role of day length changes has been controversial and in all probability different from that in the temperate zone (Chandola et al. 1983).

Coming to the realm of ultimate factors, common sense suggests that a good strategy for a breeding individual would be to reproduce at such a time that when its chicks are growing there is plenty of food available and they never run short of food supply. Proper feeding of chicks would ensure their wholesome growth and development which would contribute toward their survival. This would in turn translate into fitness benefits for the parents. Therefore, since food availability is the single most important factor governing nestling survival, the factors which govern the food cycles are also really important. Hence, the timing of these factors ultimately leads to the evolution of well-defined breeding seasons for different species of birds.

Ultimate factors themselves may have little or no value as predictive information enabling individuals to time their reproductive effort correctly. Here the proximate factors – day length for instance – come into play as they are reliable environmental cues that the individual can perceive and use as predictive information. Sometimes it may appear if birds of a particular species do not breed at the time coinciding with maximum food availability in their environment. This may be the case when tradeoffs are being made between food on the one hand and other crucial factors affecting survival of chicks on the other hand. Predation pressure may be a strong factor. Sometimes, abiotic factors such as temperature, which can strongly influence survival of nestlings, can also come into play.

## 3.2 The Timing of Nesting

Largely, inspired by ideas initially proposed by David Lack, ornithologists have built upon what has come to be known, and occupies a central position in ornithology, as the "food availability-breeding time" hypothesis (Perrins and Birkhead 1983).

Food availability as an important factor in timing of reproduction is probably more crucial for birds than it is for mammals. In mammals, milk production requires only a sufficient amount of food supply on which the adults feed but in the case of birds, which feed their young on food more or less as it is, i.e., without any processing or predigestion (with some exceptions), the food supply has to be both abundant and constant. Since peaks of food abundance may be very short, the breeding seasons of birds generally tend to be asymmetrical compared to those of mammals (Dawson 2002).

Food availability itself depends upon a number of environmental factors. In the context of India, the single most important factor influencing food cycles of birds and indeed most living organisms is the monsoon. For most terrestrial species, across India, the nesting activity is closely tied to the monsoon because they tend to feed their nestlings on insects and other invertebrates, whose abundance dramatically increases during the monsoon (Padmanabhan and Yom-Tov 2000). Since growing chicks require protein for body building, this requirement is met by live animal food. Hence even those birds, which as adults may feed on grain, nectar, or other plant matter, require an abundant supply of insects for raising their brood.

As far as the piscivorous birds are concerned, it is generally held that the monsoons trigger off their food cycles in the following manner,

Monsoon Rainfall → Plankton Bloom & Fish Spawning → Increased Fish Production (& also dispersal of fish due to flooding) → Bird Reproduction

The Painted Stork breeds after the monsoon, i.e., in the month of August till March in North India and at other times in the rest of India (Ali and Ripley 1987; Bolster 1923; Kahl 1970). Indeed, the Painted Stork and for that matter other species of colonial waterbirds too are so dependant upon the monsoon that they have been recorded to skip breeding altogether if the rains fail and that area experiences a drought (Breeden and Breeden 1982). Besides food production, the monsoon may have other desirable effects too. Due to the flooding action of the monsoon rains, the prey is dispersed far and wide and islands are formed which effectively isolate the nesting colonies from ground predators (Urfi 1998, 2010).

For those storks which are not piscivorous, the monsoon may still play a significant role in governing their nesting time. A case in point is the Asian Openbill which subsists on a diet of Apple Snails (genus *Pila*). These storks arrive in an area at the commencement of rains and this period coincides with the reemergence of snails from dry season torpor (Hancock et al. 1992).

## 3.3 Monsoon: The Key Factor

Water is so crucial for life and for countries like India the monsoon is the chief supplier of moisture by way of precipitation. Therefore, it is important to understand the dynamics of the monsoon before moving on to study the relationship between the patterns of monsoon, their impacts on food cycles, and Painted Stork nesting periods. Another good reason to have some idea about the dynamics of the monsoon is that in today's context, there is immense concern about the possible disruption in monsoon patterns due to global climate change (Qiu 2008). Therefore, the effects of the monsoon on biodiversity need to be investigated and understood so that effective conservation strategies can be developed to meet the challenges.

First of all, in monsoonal countries like India, the yearly precipitation is not uniform but highly clumped. In only some months a particular region may receive the majority of its rainfall and so all life-sustaining activities have to be timed to take advantage of this. The monsoon, a derivative of the Arabic word *mausam*, meaning season was used by Arab sailors several centuries ago. Essentially, it refers to a system of moisture-laden winds that blow from the southwest during 6 months of the year and from the opposite direction during the rest of the year (Hawkins 1986). These winds, likened to a gigantic sea breeze, are associated with differential heating of the Indian subcontinental land mass and the Indian Ocean at different times of the year. Since these winds are of two types, blowing in different directions, they are treated as two types of monsoons. One is the "southwest" or "summer" monsoon, in which from May onward the winds blow from sea to land (Fig. 3.2a). The other is the "northeast" or "winter" monsoon in which the winds blow from land to sea, after September (Fig. 3.2b). Since the latter is actually a reversal of the wind direction, it is also known as the "withdrawal" monsoon.

By May the summer monsoon starts advancing toward the Indian subcontinent and hits the coast of Kerala, in South India, by 1 June or so. The summer monsoon has two branches or arms which as they move inward in the Indian subcontinent bring rain in their wake. While one branch after hitting the Kerala coast travels northward along the western coast of India, not crossing eastward due to the barrier created by the Western Ghats all along the coast, the other branch, also known as the Bay of Bengal branch, enters the subcontinent from the east. Some of the world's rainiest spots, like Cherrapunji in the northeastern India receive early rainfall. As the monsoon moves westward, across the Indo-Gangetic plains of North India, it makes rapid progress, reaching Delhi by end June or thereabouts (Fig. 3.2a).

The summer monsoon, though the major monsoon system for much of the Indian subcontinent, does not bring rainfall uniformly all across the subcontinent and so there is a huge disparity in the amount of rainfall received across the country. Some areas, mostly in the north, receive most of their annual rainfall in the months June–September while other areas, particularly those in the central and southern India, receive scanty or no rainfall during the same period. So, while, over 70% of the annual rainfall over India as a whole and around 90% of the annual rainfall in certain localities are brought by the summer monsoon, the withdrawal monsoon brings rainfall in other areas in other months. In some southern sites, like Chennai, close to which many important heronries are located, much of the rainfall is due to the winter monsoon (Fig. 3.3b).

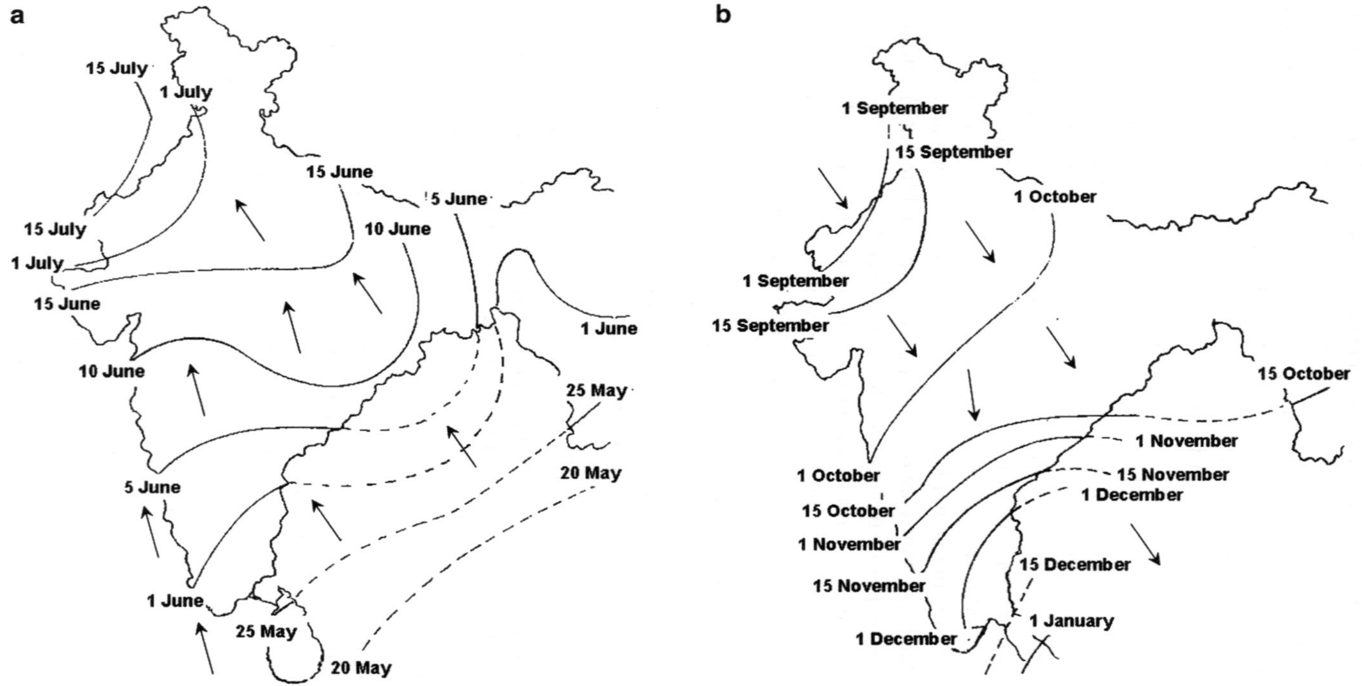


**Fig. 3.2** The two major Monsoon patterns of India. (**a**) Summer monsoon and (**b**) winter monsoon. *Arrows* denote the direction of monsoon winds. Note in the case of the summer monsoon the direction of winds is from sea to land and in the winter monsoon the wind direction is reverse, from land to sea. Redrawn from the Encyclopedia of Indian Natural History (Hawkins 1986)

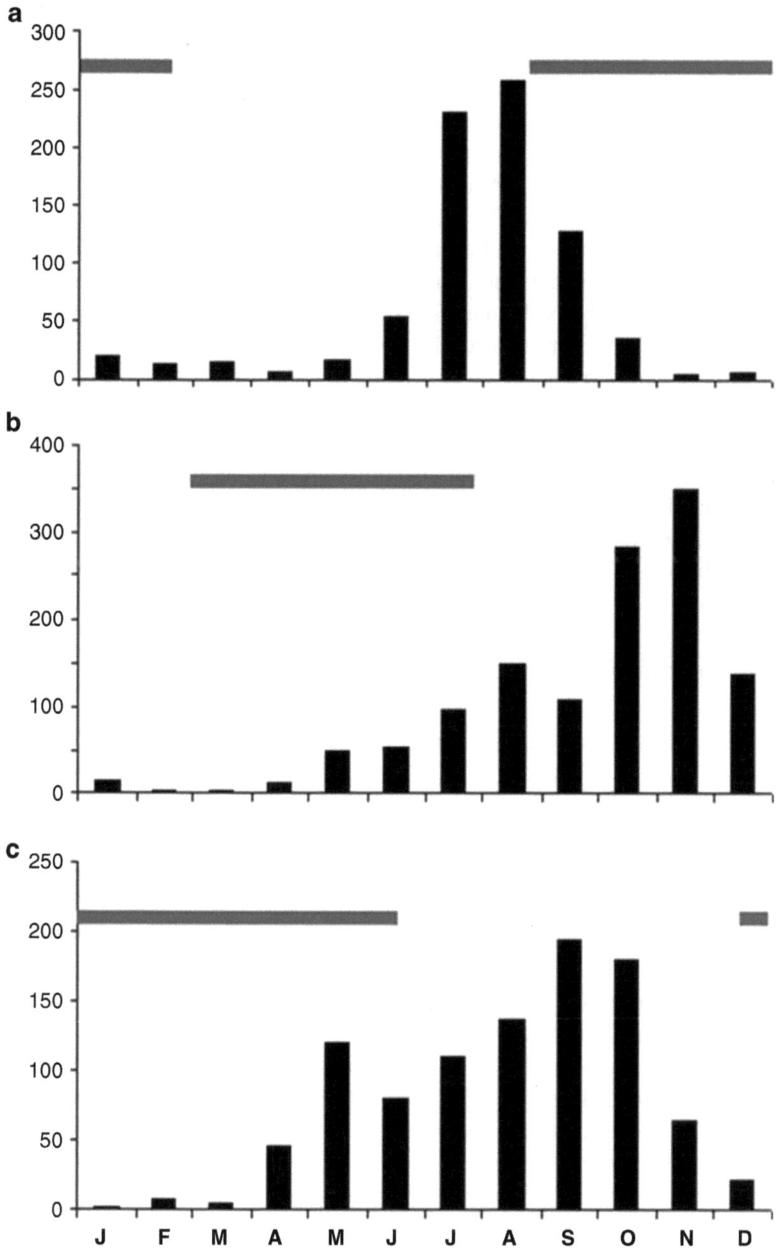


**Fig. 3.3** The differences in rainfall profiles of three Indian cities. (**a**) Delhi, (**b**) Chennai, and (**c**) Bengaluru. Horizontal axis months. Vertical axis mean total rainfall for each month in mm. The *horizontal bar* shows the nesting season of the Painted Stork as follows: (**a**) Delhi Zoo, (**b**) Kalpakkam, and (**c**) Kokkare Bellur. Site details are given in Table 3.1. Weather data from (WMO) World Meteorological Organization (2010)

## 3.4 Variability of Painted Stork Nesting Dates Across India

The causal linkages between bird reproduction and monsoon patterns from India have been relatively unexplored in context of Painted Stork although Ali and Ripley (1987) observed that it breeds in August to October in North India, November to March in the South India, and March to April in Sri Lanka. This clearly suggests that in India there are two breeding zones. The picture gets clearer if we plot the nesting dates of Painted Stork on a map. For such a study the most crucial information is nest initiation date, which is available from a number of sites across India. On the basis of its nesting period basically the Painted Stork population of India gets grouped into two clusters (Fig. 3.4). The first cluster, located in the north, is a broad area stretching from Delhi in the north up to western Gujarat. At all the sites in this cluster, the Painted Stork nesting dates are August–September, stretching to October in some cases. However, in the second cluster, lying in Central and South India the nesting dates are not only more variable but also different. They are typically in the months January–March, though in some cases even as early as December (Table 3.1). In both the clusters, the monsoon patterns are different and the nesting dates seem to be correlated with them. However, more studies are required for more clarity on this issue.

### *3.4.1 Correlations with Monsoon Phenology*

There is some evidence that Painted Stork nesting is influenced by both monsoon strength as well as phenology. At both Sultanpur (Urfi et al. 2007) and Delhi Zoo (Urfi 1997), a weak correlation was observed between the total yearly rainfall and the number of nests built. Earlier studies at the zoo (Desai et al. 1977) also suggested a correlation between monsoon performance and reproductive success, which is discussed in the next section.

In North India, it seems that generally the Painted Stork arrival at the breeding site is approximately a month or so after the first monsoon showers in that particular region. At Sultanpur Bird Sanctuary, the average arrival date of the first Painted Stork at Sultanpur during 2003–2005 was 18 August and the first eggs were laid between 17 August and 23 September (Table 3.2). This suggests that after arrival it takes, on an average, about 10–12 days for storks to complete courtship, nest building, etc.

There is of course considerable yearly variation in the time when the monsoon actually arrives in any locality. Some records of Painted Stork arrival dates have been maintained by the Delhi Zoo staff, chiefly for the period 1990–2000, which were made available to us at the time of writing this chapter (A.K. Malhotra, personal communication). Although one does not know how accurately the arrival dates were recorded and whether the same criteria were adhered to each year, plotting arrival dates of monsoon in the city of Delhi and arrival dates of Painted Stork in the Delhi Zoo, a positive relationship between the two is apparent (Fig. 3.5). If a long-term heronry monitoring program is in place at the zoo, with all the parameters, including arrival dates recorded accurately, then it may be possible to study the

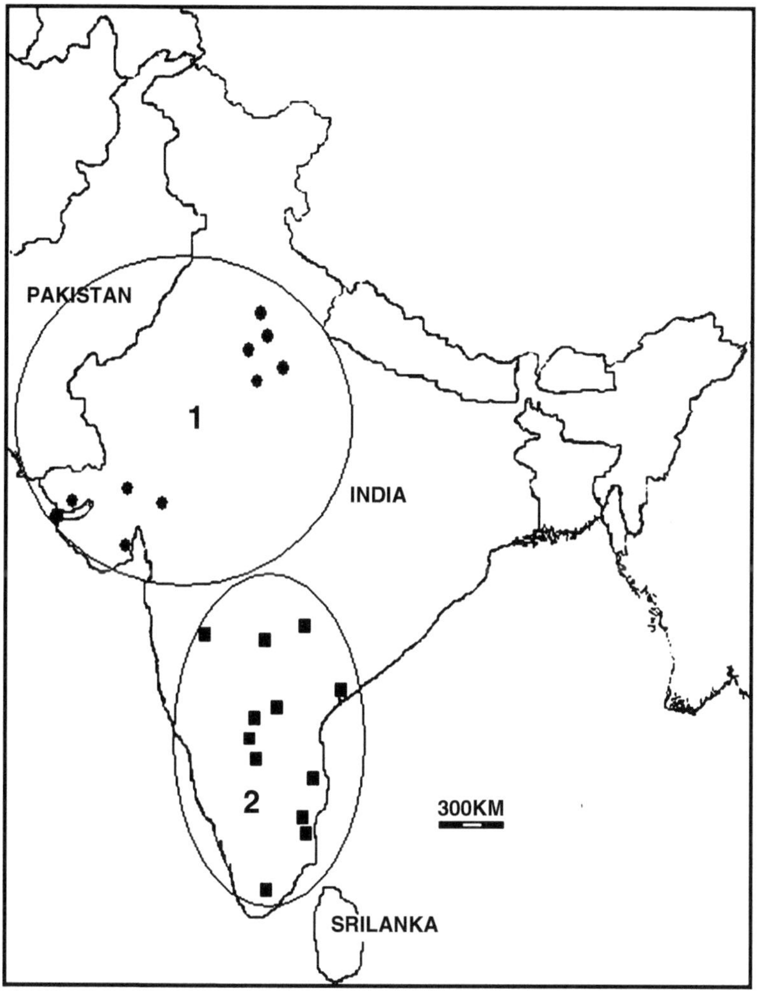


**Fig. 3.4** Variations in nesting dates of Painted Stork from different sites across India. The details for each site are given in Table 3.1. In zone 1 *filled circles* denote nest initiation in August–September and in zone 2 *filled squares* denote nest initiation in the months December–January (or even later in some cases)

relationship between the monsoon patterns and Painted Stork nesting in a more rigorous manner, as discussed in Chap. 8.

### *3.4.2 Nesting Success*

Although nesting success (NS) is an important fitness parameter, huge controversies have surrounded its estimation. The standard practice was to express NS as number

**Table 3.1** Variability in nesting dates of Painted Stork across India[a]

| Site and location | Nesting period | References |
|---|---|---|
| North India | | |
| Delhi Zoo, Delhi | Mid August | Urfi (1997) |
| Sultanpur, Haryana | Mid August | Urfi et al. (2007) |
| Bijana, Haryana | Mid August | Urfi (1993) |
| Govardhan, Uttar Pradesh | September | Hume and Oates (1890) |
| Keoladeo Ghana, Rajasthan | End August | Ali and Vijayan (1983) |
| Man Marodi Island, Gujarat | End August | Urfi (2003a) |
| Seelaj, Gujarat | End August | Urfi (2003b) |
| Bhavnagar City Gardens, Gujarat | End August | Parasharya and Naik (1990) |
| Munjasar Tank, Gujarat | Begin September | Tere (2008) |
| Chikhaliya, Gujarat | September | Anonymous (2008) |
| Central and South India | | |
| Bhadalwadi, Maharashtra | Last week March | Pande (2006) |
| Karanji Tank, Karnataka | Mid February to May/June | Jamgaonkar et al. (1994) |
| Ranganthitoo, Karnataka | Mid February to May/June | Jamgaonkar et al. (1994) |
| Kokkare Bellur, Karnataka | December/January to June | Neginhal (1977), Saxena (1980) |
| Kaggaladu, Karnataka | November to May | Pai (2009) |
| Veerapura, Andhra Pradesh | Last week January | Bhat et al. (1991) |
| Chintapalli, Andhra Pradesh | January to June/July | Pattanaik et al. (2008) |
| Chinna Maduru, Andhra Pradesh | January to June/July | Pattanaik et al. (2008) |
| Telineelapuram, Andhra Pradesh | Mid May | Nagulu and Rao (1983) |
| Vedurupattu, Andhra Pradesh | November/December–May | Sharma and Raghavaiah (2000) |
| Koonthangulam, Tamil Nadu | End March/April (also end December ) | Rhenius (1907), Webb-Peploe (1945), and Wilkinson (1961) |
| Kalpakkam, Tamil Nadu | January/February | Rajaram (1999) |
| Vedanthangal, Tamil Nadu | Early January | Paulraj and Gunasekaran (1988) |

[a]In those cases where nesting dates were casually mentioned, nesting period was estimated from age or size of the nestlings (see Chap. 5), if such information was available in reports. For other details see text

**Table 3.2** Nesting details of Painted Stork at Sultanpur National Park during 2002–2005

| Year | First arrival | First egg | Last eggs | No. colonies | No nests |
|---|---|---|---|---|---|
| 2002 | – | – | – | 1 | 26 |
| 2003 | 20 August | 1 September | 5 November | 5 | 96 |
| 2004 | 28 August | 12 September | 17 October | 1 | 57 |
| 2005 | 8 August | 17 August | 8 November | 2 | 79 |

*Source*: Urfi et al. (2007)

of fledged as a proportion of total eggs counted. But as was pointed out (Mayfield 1961, 1975) by this method, the number of nests that failed before the investigator started his study was totally disregarded. Consequently, the result would be a gross overestimate of the actual, or close to real NS in the population. Modern approaches, using the Mayfield method, define NS as the probability of a nest surviving to produce at least one young to a given age of interest (Frederick and Spalding 1994).

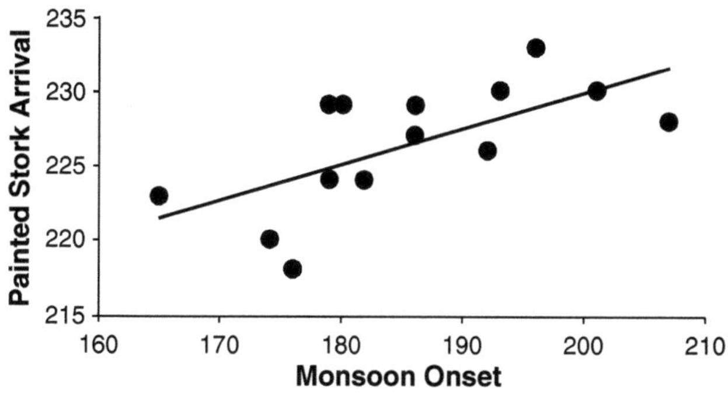


**Fig. 3.5** The relationship between monsoon onset dates in Delhi and the arrival dates of Painted Stork at the Delhi Zoo expressed as number of days since 1 January. Based on data provided by the Delhi Zoo

Since, of late, no studies on NS have been published, we therefore have to rely upon the earlier study of Desai et al. (1977) who while using the traditional method estimated NS to lie within the range 34–54% for the period 1966–1971. A broad correlation was observed between years of high rainfall and NS. For instance, the years 1967 and 1969 were both years of good monsoon and these were also years when some of the highest values of NS, 54 and 42% respectively, were recorded. The poorest rainfall was received in 1968 which was also a year of low NS (34%).

## 3.5 Courtship Patterns

In August, when they arrive at the Delhi Zoo, Painted Storks have a fresh glistening look and a spotless white plumage. Their bill looks bright waxy yellow and the head is fully bald (see Chap. 5). However, as time passes the plumage gets soiled and dirty.

The courtship patterns of Painted Stork have been studied in detail by Kahl (1971a) using cinematography and replaying the footage at different speeds. Accordingly, upon arrival at the colony site the birds at first simply stand on tree limbs in bachelor groups of unattached males and females. Then adult males attempt to take possession of a suitable territory, which could be a nest of yesteryear or in exceptional cases a new nest site. Sometimes, they could also be nests of other species like Asian Openbill.

A male Painted Stork attracts a female by advertising his presence, with such subtlety and finesse that his actions hardly look like a proper display. A lone male may appear to be doing nothing except preening himself and rearranging twigs in the nest but he may be actually displaying (Kahl 1971a). Should another male encroach upon an occupied territory, he is rudely driven off by a powerful jab of the bill. Some interlopers fight back, and when conflicts escalate, the intruder may even succeed in usurping the territory. A female, drawn to a displaying male, is initially attacked, as if she were a rival male. But instead of retaliating, she adopts a meek and submissive posture, her head held low, wings spread wide, and the bill gaping. The male soon recognizes her as a potential mate and if a pair bond develops between them, they immediately embark upon a hectic schedule of nest building and mating.

### 3.5.1 Age at First Breeding

Being large-sized birds, the age at first reproduction in the Painted Stork could be expected to be no less than a couple of years. The only authentic information on record is from some color-marked birds at the Delhi Zoo colony. In 1968, ten nestlings of known ages were color marked and subsequently released into the wild in February 1970. However, the birds became tame and did not leave the zoo premises. In September 1972, three of these color-marked birds paired and mated with the wild storks in the zoo colony from which it can be deduced that the Painted Stork becomes sexually mature and breeds for the first time at the age of 4 years (Shah and Desai 1975).

### 3.5.2 Nest Building

The nests of Painted Stork are a jumble of twigs of varying sizes placed mostly in the crowns of trees. Essentially, they are platform nests. At the now extinct colony of Goverdhan, in North India, Painted Stork nests measured as much as 2 ft in diameter and weighed from 2.2 to 2.7 kg (Hume and Oates 1890). Desai et al. (1977) examined 15 nests at the Delhi Zoo and recorded their internal and external diameters and their depth, and the time taken for the construction of each. Nest construction time ranged from 4 to 8 days. Expectedly, nest construction time and nest size were positively correlated (Fig. 3.6).

At the Delhi Zoo, we observed (Meganathan and Urfi 2009) that nests built from a scratch were rare ($n < 10$) as previous years nests or those built by other birds were generally utilized. The adding of fresh leafy twigs of Neem and Mesquite was commonly observed.

Burger (1985) examined the patterns of habitat selection among different species of colonial waterbirds in a mixed species colony and found interesting patterns, for instance, vertical stratification of nests with respect to body size of birds belonging to different species. In the case of mixed species heronries in India while a similar pattern has not been recorded, large birds like the Painted Stork have always been observed to build nests on the outer canopy or on the surface, probably because they need an open space for take off and landing. Their movements would be considerably restricted if the nest was placed in the inner branches of the tree. Most of the smaller species like cormorant, egret, heron, etc. however, build their nests in the interior of the trees, firmly wedged between some of the smaller branches. Interestingly, Painted Stork nests are to be seen as low as the water mark itself in many places like the Delhi Zoo.

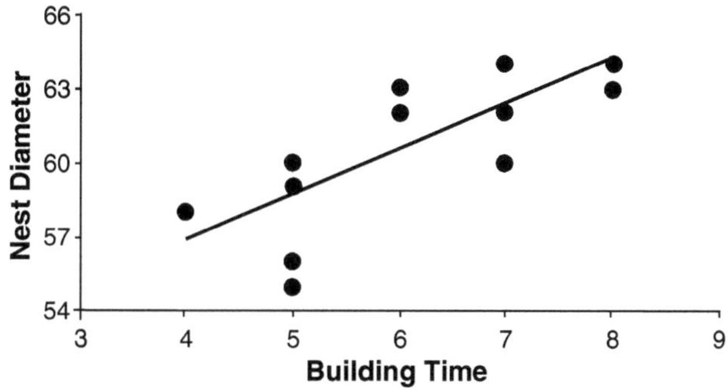


**Fig. 3.6** Relationship between outer nest diameter (cm) and number of days taken to build the nest (building time) observed for 15 Painted Stork nests at the Delhi Zoo during 1966–1971 ($r^2=0.62$). *Source*: Desai et al. (1977)

An interesting aspect of habitat selection in mixed species colonies, though not formally tested, is that while the Black-headed Ibises tend to build nests in close proximity to each other (Ali and Ripley 1987), Painted Stork nests are spaced out on the canopy surface. It also appears that nests on the canopy surface are clustered in a particular direction, though this also has not been formally tested.

## 3.6 Aggression at Colonies and Resource Partitioning

Colonial waterbirds are quite aggressive during territorial displays and in the beginning of the breeding season there is generally much action to be seen. Even birds like ibis, which appears so calm and peaceful, are recorded to inflict powerful injuries by their bills on rivals, often leading to blood shedding and sometimes even body impairment (Bildstein 1993).

An interesting question is, what are the types of inter- and intraspecific interactions taking place in mixed species colonies and are they of any significance in terms of impacting the nesting time of different species? This is also a question about resources and how they are utilized. In other words, how do species, some of which utilize similar resources, coexist in a mixed heronry?

At the Delhi Zoo, we recorded interactions per 15 min on individual Painted Storks. Very few interactions (<5%) were interspecific. The majority of interactions were territorial disputes or fights over nesting material (A.J. Urfi and T. Meganathan, unpublished data). Hence, interspecific competition is probably not likely to influence nesting time. Also, there is generally a segregation, in a spatio-temporal sense, if we take into account those species which can possibly compete with the Painted Stork at the nesting site (Fig. 3.7). Two such species are Asian Openbill and Black-headed Ibis, both being large sized and having a tendency to build nests on the same trees as the Painted Stork.

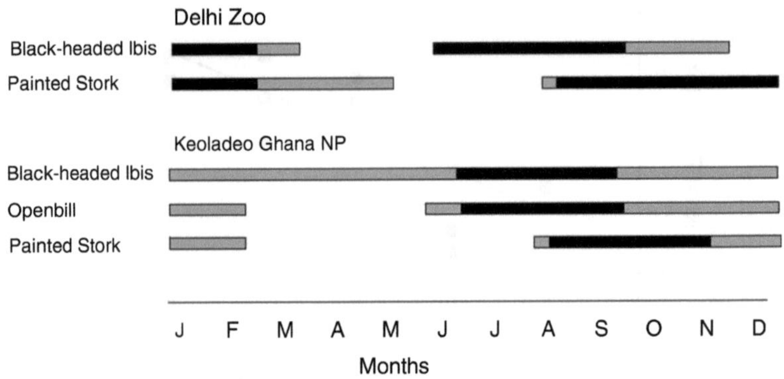


**Fig. 3.7** Figure shows how different species of similarly sized colonial waterbirds can coexist due to variations in their nesting period. *Shaded areas* denote present at the site while filled (*black*) areas denote nesting activity. Data from Sankhala (1990) and Urfi (1997)

## 3.7 Nest Substrates

Painted Stork nests are fairly large and are placed on the outer surface of the tree canopy. There is no attempt to camouflage and also no protection from aerial predators (raptors). At a gross level, it seems that selection is largely in terms of avoiding ground level predators. From a survey of several types of heronries across India, it seems that the general rule is, if the nesting trees are low (say *Prosopis* or *Acacia* clumps) they are surrounded by water, which gathered during the monsoon and now provides some sort of a barrier against ground predators. If nests are not located on islands, then large trees (*Ficus* or *Tamarindus*) are selected. Such trees may be in villages along with human settlements (e.g., Kokkare Bellur) or in agricultural fields, such as the *Ficus* tree near the village Bijana in North India (Urfi 1993). Nests on such trees are likely to be out of reach of ground predators. Certainly, in some situations nests are placed very close to the ground, as in *Euphorbia* stands at Man Marodi Island (Urfi 2003a) but because of the barrier of sea they are safe from mammalian predators.

## 3.8 Typical Behaviors at Nest

At the time of nest building, Painted Storks are observed carrying fresh leafy twigs for lining their nests, which are usually the already used nests of other birds or their own nests from the previous year, now reduced to a loose mass of dried twigs. At the Delhi Zoo they have been observed using green branches of mesquite or *Acacia* by pulling them off from trees by their bills, but at other sites, other types of vegetation have been observed (Krishnan 1980). However, Painted Storks are not unique in lining their nests with fresh, green leafy branches as many other species of colonial waterbirds, including other storks as well as egrets and herons have been recorded doing the same. Given that this is such a common behavior among colonially nesting waterbirds, the question is what purpose does it serve?

Rodgers et al. (1988) studied the use and function of green nest material by Wood Stork in North and Central Florida. They examined a number of hypothesis, such as, the role of green vegetation in concealment of eggs or young nestlings from predators, chiefly Raccoon and Fish Crow, and providing shade. However, nests are usually not left unguarded as one adult stork is generally present at the nest which has eggs or hatchlings. Also the nest contents are seldom covered by greenery material so shading does not appear to be so important, as it is perhaps for some other species. The Wood Stork, like the Painted Stork, provides shade to nestlings by stretching out their wings in a delta pose during warm periods of the day.

A possible role for the greenery material could be in terms of aiding nest sanitation since nestlings regularly defecate in the nest. However, since in the Wood Stork greenery deposition is initiated prior to egg deposition and declines subsequently, its sanitation role if any is likely to be quite limited. Nest greenery actually tends to prevent guano from passing through the nest twigs and this in fact reduces the plausibility of the sanitation hypothesis with respect to guano.

In the case of some species of raptors, it has been hypothesized that greenery use aids in repelling ectoparasites via release of secondary compounds during drying or decay of plant material. European Starlings are reported to choose plants for nest building whose volatile compounds are likely to inhibit arthropod hatching (but not adult stage mites) and bacterial growth. Rodgers et al.'s study suggests that some of the vegetation used by Wood Stork does support this hypothesis, especially the aromatic or resinous species such as cypress, red cedar, poison ivy, Brazilian pepperbush, pine species, and wax myrtle. However, Wood Stork nests have a heavy infestation by dermestids larvae, biting lice, and mites, against which the vegetation used by storks has no effect. Significantly, in the different colonies studied by Rogers and coworkers, the use of vegetation was more a reflection of what was easily available in any particular locality.

Possibly, green material in nest lining may serve more than one function. For instance while one of its advantages could be that it may help in retaining food that is regurgitated by parents, especially in the case of food boluses, it may also have a role to play in terms of maintaining the pair bond as well as the structural integrity of the nest. Through a series of experiments Rodgers et al. showed that the green material used for nest lining could have a role to play in insulating the nest and reducing the energetic cost of incubation. Greenery itself or in combination with guano may help in plugging holes in a nest which to start with has a twiggy, porous structure.

Some other typical stork behaviors exhibited by Painted Stork have to do with countering excess heat in their environment. Unlike some other birds, storks do not indulge in gular fluttering (del Hoyo et al. 1992). However, to bring down body temperatures they use different techniques, one of which is "urohydrosis,"

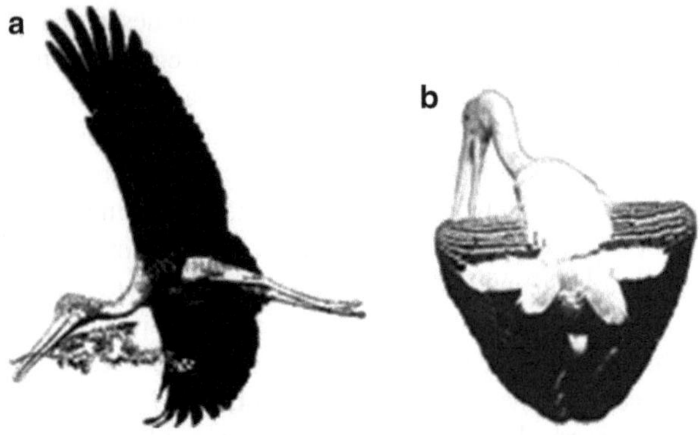


**Fig. 3.8** Typical stork behaviors at the time of nesting. (**a**) Painted Stork carrying fresh leafy twigs for nest lining, (**b**) wings folded in the delta pose to shield young for direct sunlight

i.e., defecating directly onto the legs, thus enabling heat loss due to evaporation (Kahl 1963). Another typical stork behavior observed at nest is that of shielding the young from direct sunlight by assuming the delta wing pose, i.e., spreading the wings half-open but bent at the wrists in the form of a shield (Fig. 3.8b). Much has been written about the spread-wing postures of storks (Kahl 1971b) and in any case the same pose which is used to shield nestlings from direct sunlight is also used for sunning, which is helpful in restoring the curve to feathers after soaring has left them misshapen (del Hoyo et al. 1992).

Since the sun shines upon the nest from different angles during the course of a day, it is held that parent birds shift their position accordingly. This idea has not been tested, though it is recorded in anecdotes (Abraham 1973),

> The villagers assure me that the parent birds stand on eastern side of the nest in the forenoon and on western side of the nest in the afternoon to protect their young ones from the sun…. However, even the casual visitor cannot but notice the way the Painted Stork protects the young ones from the sun by spreading its broad and long wings over the young ones as a sort of improvised roof.

### *3.8.1 Nest Parasites*

Little information seems to be available about nest parasites of heronry birds from India, although one comes across references on ectoparasites whose hosts are storks (Clay and Hopkins 1960). From the work of Rodgers et al. (1988), a general idea can be had about nest parasites in nests of *Mycteria*, but in the Indian context clearly there is a need for more studies about this aspect.

## 3.9 Nonbreeding Birds at Nests

Of all the Painted Storks that initially congregate at a nesting colony not all eventually breed. At Delhi Zoo and Sultanpur, it has been observed that approximately a quarter of the adult population emigrates out of the area within the first few weeks of arrival. The ones left behind are the breeding individuals, i.e., those birds who have successfully contested territories, along with some nonbreeding individuals who simply hang around. Let's say, if the total number of nests built at a site is 50 then assuming two individuals per nest, the expected adult count should be around 100. However, careful counts (Urfi et al. 2007) have revealed that there is generally a small surplus, of about 20–30%, which cannot be accounted for on the basis of a nest count alone. Since there have not been any studies using ringed or marked birds, we do not know anything about these surplus birds.

At present it is difficult to account for these apparently surplus birds. In many colonial waterbirds, the nesting site also serves as a roosting site, so the occurrence of additional individuals may be coincidental or related to opportunities for obtaining food while chicks in the colony are being fed (Burger 1981). However, Painted Storks congregate only during the nesting period, so this explanation is unlikely. They could be immature adults, probably <4 years or adults attempting to seek extra-pair copulations (Westneat et al. 1990). Helping behavior has been observed in several birds (Skutch 1976). So, perhaps, some of these apparently extra individuals are genetically related to the breeding pair and hang around the nest as helpers. In seabirds, nonbreeding individuals, both immature and adult, may be permanent residents within a colony, defending territories, constructing rudimentary nests, and even assisting in raising young (Kharitonov and Siegel-Causey 1988).

Cases of polygyny have been recorded in some Ardeids. In one study polygynous trios in 7.7% of Cattle Egret nests were recorded (Fujioka 1986). Several cases of polygyny were observed in the population of Asian Openbill nesting at Raiganj Wildlife Sanctuary in India (Datta and Pal 1995). Thus, storks which have always been found to breed monogamously (Ali and Ripley 1987; Coulter et al. 1989) may actually be more polygamous than previously believed. In the case of Painted Stork, it is difficult to account for the surplus individuals at nesting colonies, and clearly more studies are required.

## 3.10 Factors Causing Nest Losses

Losses at egg and nestling stages significantly impact fitness in birds and they can be broadly classed under the categories of biotic and abiotic factors. In the context of Painted Stork, the role of biotic factors, which chiefly include predation, infertile eggs, and starvation and abiotic factors, which include weather, human disturbance, and related factors, has been recently reviewed (Urfi 2010), an outline of which is provided here.

### 3.10.1 Predation on Eggs and Nestlings

Nest predation is an important ecological factor influencing reproductive success in birds. Ensuring safety from ground predators, chiefly mammals, has been a strong selective force in the evolution of coloniality in birds (Brown and Brown 2001). Although most Painted Stork colonies are located out of harm's reach, there are also instances on record when mammalian predators have managed to reach island colonies by swimming, when the water is shallow or a bridge is formed due to a sudden drop in water levels. Pande (2006) records an instance at Bhadalwadi Tank of stray dogs, Common Mongoose, Jackal, and Wolf being able to gain access to Painted Stork nests when the water level surrounding the colony dropped unexpectedly.

Although cases of mammalian predators reaching nesting colonies located on islands off the sea coast by swimming are probably rare, at Man Marodi Island in the Gulf of Kutch, where Painted Storks build nests on *Salvadora*, quite close to ground level, jackals are known to attack nestlings (Urfi 2003a). Reportedly, they arrive on the island by swimming at low tide, from nearby mainland areas. At Ranganthittu Bird Sanctuary, troops of Bonnet Macaque were recorded to swim across the river to the bird colonies on islets and plunder the eggs in the nests (Neginhal 1982). At Delhi Zoo, Rhesus Macaques were observed to cause much damage to Painted Stork nesting in Pond 1 by romping around on the trees (A.J. Urfi and T. Meganathan, unpublished observations).

Village heronries such as Kokkare Bellur or heronries on large trees in city parks, such as gardens in Bhavnagar city, are good examples of safety from ground predators being ensured due to location of nests at a height. The only way in which ground predators can have access to nestlings is when they accidentally fall off from their nests. While the predators in village heronries are mostly feral dogs, (Subramanya and Manu 1996) in island colonies they may be the Mugger Crocodile, as in the case of Ranganthittu (Neginhal 1982).

Since most observations on nest losses are based on observations made during the day, the impact of night-time predation remains largely unknown. However, some scattered reports confirm its occurrence. For instance, Common Indian Monitor climbing on trees and devouring the eggs of Painted Stork in evenings at the Delhi Zoo is on record (Meganathan and Urfi 2009). Interestingly, Malayan Water Monitor Lizard has been reported to devour eggs and nestlings of Milky Stork in South Sumatra province of Indonesia (Iqbal et al. 2008). At Keoladeo Ghana, most kills of fledgeling Painted Stork by eagles (*Aquila* sp.) were recorded on moonlit nights (Naoroji 1990), indicating that much predation on heronry bird's eggs and nestlings by birds happens at night.

The main predation pressure is of course exerted by aerial predators, i.e., raptors, against which there is little protection. Several points are of interest in this context. First, in North India, the period when Painted Stork nests have fledglings is the same time when the influx of migratory raptors begins (Naoroji 1990). Several species of *Aquila* and other raptors overwinter in the Indian subcontinent (Ali and Ripley 1987) and their migratory influx starts in September. Second, recent studies at Delhi

Zoo and Sultanpur (Urfi et al. 2007; Meganathan and Urfi 2009) have hinted at a broad correlation between the body sizes of predator and prey. For instance, while House Crows attack small nestlings (<15 days old), Black Kites, which are considerably larger in size, were observed to take older chicks at the Delhi Zoo. Third, there are differences in predator species at colonies located in urban areas and those in the country, as would be expected. While at Delhi Zoo, which is located in a populous city, omnivorous, feral birds like crows and kites account for most of egg and nestling losses, at natural areas like Sultanpur and Keoladeo, those raptors which are partial to undisturbed areas in the country such as Greater Spotted Eagle, Steppe Eagle, Imperial Eagle, and Pallas's Fishing Eagle are recorded to be the main predatory agents (Naoroji 1990; Urfi et al. 2007). This therefore leads to the question, since at urban sites predation pressure is lower, compared to colonies in the open countryside, could this be an additional inducement for the formation of colonies in urban premises, besides conditions of safety and availability of suitable nesting substrates?

Detailed observations on the mode of attack by *Aquila* sp. are known largely through the observations of Naoroji (1990) at Keoladeo Ghana, Bharatpur. For instance, only nestlings were taken and adults were seldom attacked. The hunting method of raptors was opportunistic and cases of their trying to bully adults, mostly unsuccessfully, to leave nests were also recorded. Kleptoparasitism among the raptors and often involving both House Crow as well as Jungle Crow was common. An examination of nestlings attacked by raptors revealed that a number of individuals had sustained head and neck injuries, suggesting that most attacks were directed toward the head. Earlier, Lowther (1949) recorded a breast portion of Painted Stork eaten and the rest discarded. Interestingly, while at the Keoladeo, nests in isolated patches were observed to be predated as frequently as nests in groups, spatial variations in predation intensity were observed at Delhi Zoo (Meganathan and Urfi 2009).

### 3.10.2 Infertile Eggs

Eggs lying in nests, generally untouched by predators and hence assumed to be infertile, have been recorded at Delhi Zoo and Sultanpur (Desai et al. 1977; Meganathan and Urfi 2009; Urfi et al. 2007).

### 3.10.3 Starvation

Starvation is often attributed to be a major cause of nestling loss in birds, especially in the first few weeks after hatching. At the Delhi Zoo the figure of yearly starvation deaths was estimated at around 38% although the deceased nestlings were not examined to study body condition and to verify the cause of death (Desai et al. 1977). In some years, notably 1966, 1967, and 1971 the number of nestling deaths assumed to have been caused due to starvation was recorded to be 44, 50, and 55% respectively. Compared to the years 1968–1970 when such mortalities were below 33%, this is a high number and could be due to shortage of food and correlated to the yearly monsoon performance. Competition between the siblings, in which older nestlings monopolized all the food regurgitated by the parents on the nest floor, resulting in the younger siblings losing condition and eventually dying have been alluded to in the above report.

### 3.10.4 Weather and Other Factors

Fluctuations in environmental temperatures leading to nestling mortalities in European White Stork have been reported (Jovani and Tella 2004). The fact that Painted Stork exhibits a wing spread behavior at nest to shield nestlings from the sun and also sometimes regurgitates water to bring down nest temperatures on hot days suggests that heat issues are significant. Depending upon where a colony is located, environmental temperatures are significant accordingly. In the Delhi region for instance, where nesting starts in end August or early September, environmental temperatures start dropping by November as winter approaches (Fig. 3.9a). Although, in India, the winter is not so severe (except in the Himalayas) as it is in the temperate zone, in December and early January temperatures below 4°C are also sometimes recorded from Delhi. Compared to sites lying further south where the minimum temperatures do not drop so low (Fig. 3.9b, c), the winter in Delhi and northern sites in general would be considered quite severe in relative terms. Around this time of the year some mortalities of Painted Stork have been recorded from Delhi. For instance, bodies of juveniles and adults (<5) were recorded immediately succeeding very cold days during 1988–1992 at the Delhi Zoo (A.J. Urfi, unpublished observations). However, since the corpses observed to be strewn on branches of trees close to the nests were not recovered for a postmortem examination, it could not be ascertained if these deaths were indeed due to hypothermia. Dead nestlings and adults were also observed during the study in 2005–2006 (Meganathan and Urfi 2009). In South India, which has a much more equable climate, environmental temperatures never drop as low as at North Indian sites (Fig. 3.9b, c).

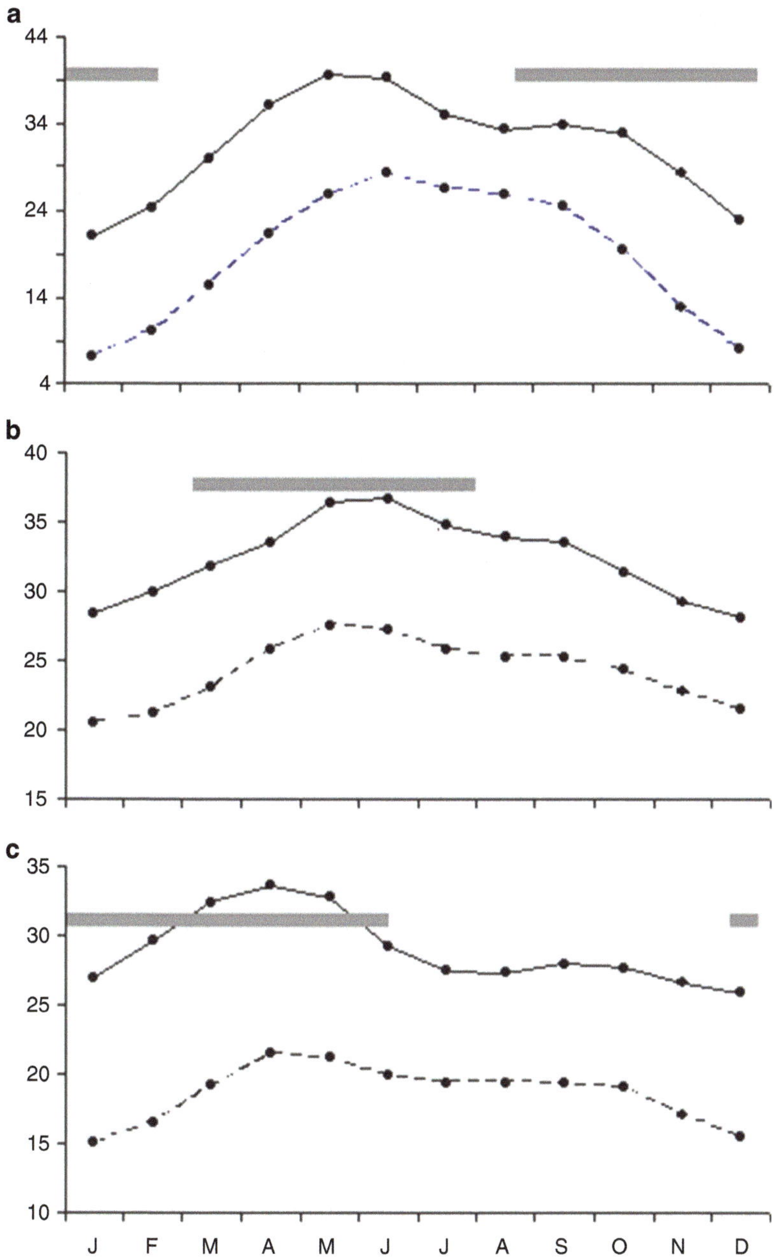


**Fig. 3.9** Maximum (*upper solid line*) and minimum (*lower dashed line*) mean temperatures of three Indian cities, (**a**) Delhi, (**b**) Chennai, and (**c**) Bengaluru. Horizontal axis months. Vertical axis mean temperatures in °C. Note scale differences in the vertical axis. The *horizontal bar* shows the nesting season of the Painted Stork as follows: (**a**) Delhi Zoo, (**b**) Kalpakkam, and (**c**) Kokkare Bellur. Weather data from (WMO) World Meteorological Organization (2010)

# References

Abraham S (1973) The Kanjirankulam breeding bird sanctuary in the Ramnad district of Tamil Nadu. J Bombay Nat Hist Soc 70:549–552
Ali S, Ripley SD (1987) Handbook of the birds of India and Pakistan together with those of Bangladesh, Nepal, Bhutan and Sri Lanka. Oxford University Press, Delhi
Ali S, Vijayan VS (1983) Hydrobiological (ecological) research at Keoladeo National Park, Bharatpur (first interim report). Bombay Natural History Society, Mumbai
Anonymous (2008) BCSG events. Painted Stork and Sarus watch. Flamingo 6:17
Bhat HR, Jacob PG, Jamgaonkar AV (1991) Observations on a breeding colony of Painted Stork *Mycteria leucocephala* (Pennant) in Anantpur district, Andhra Pradesh. J Bombay Nat Hist Soc 88:443–445
Bildstein KL (1993) White ibis. Wetland wanderer. Smithsonian Institution Press, Washington
Bolster RC (1923) Notes on the breeding season of the Painted Stork (*Psedotantalus leucocephalus*). J Bombay Nat Hist Soc 29:561–562
Breeden S, Breeden B (1982) The drought of 1979–1980 at the Keoladeo Ghana Sanctuary, Bharatpur, Rajasthan. J Bombay Nat Hist Soc 79:1–37
Brown CR, Brown MB (2001) Avian coloniality, progress and problems. In: Nolan V Jr et al (eds) Current ornithology. Kluwer, New York
Burger J (1981) A model for the evolution of mixed-species colonies of ciconiiformes. Quart Rev Biol 56:143–167
Burger J (1985) Habitat selection in temperate marsh-nesting birds. In: Cody ML (ed) Habitat selection in birds. Academic, New York
Chandola A, Bisht M, Bhatt D (1983) Reproductive strategies in birds of the tropics. In: Margaris NS, Faraggitaki MA, Reiter RJ (eds) Adaptation to terrestrial environment. Plenum Press, New York
Clay T, Hopkins GHE (1960) The early literature on mallophaga (part 4, 1787–1818). Bulletin of the British Museum. Entomology 9:1–61. http://www.archive.org/stream/bulletinofnatura09entolond#page/10/mode/2up/search/Tantalus. Accessed 12 Nov 2010
Coulter MC, Balzano S, Johnson RE, King CE, Shannon PW (1989) Conservation and captive management of stork. In: Proceedings of an international workshop at the New York's Zoological Society's Wildlife Survival Center, Saint Catherine's Island. Stork Interest Group, Georgia
Datta T, Pal BC (1995) Polygyny in the Asian Openbill (*Anastomus oscitans*). Auk 112:257–260
Dawson A (2002) Photoperiodic control of the annual cycle in birds and comparisons with mammals. Ardea 90:355–367
del Hoyo J, Elliott A, Sargatal J (eds) (1992) Handbook of the birds of the world, vol 1. Lynx Edicions, Barcelona
Desai JH, Menon GK, Shah RV (1977) Studies on the reproductive pattern of the Painted Stork, *Ibis leucocephalus*, Pennant. Pavo 15:1–32
Frederick PC, Spalding MG (1994) Factors affecting reproductive success of wading birds (Ciconiiformes) in the Everglades ecosystem. In: Davis S, Ogden J (eds) Everglades: the ecosystem and its restoration. St. Lucie Press, Florida
Fujioka M (1986) Two cases of bigyny in the Cattle Egret *Bubulcus ibis*. Ibis 128:419–422
Hancock JA, Kushlan JA, Kahl MP (1992) Storks, ibises and spoonbills of the world. Academic, London
Hawkins RE (ed) (1986) Encyclopedia of Indian natural history. Oxford University Press, Mumbai, India
Hume AO, Oates EW (1890) The nests and eggs of Indian birds, vol III. Taylor & Francis, London
Iqbal M, Ridwan A, Herman (2008) Local people's perspective for Milky Stork: a case from South Sumatra. Indonesia J Wetlands Ecol 1:7–8
Jamgaonkar AV, Jacob PG, Rajagopal SR, Bhat HR (1994) Records of new breeding colonies of Painted Stork *Mycteria leucocephala* in Karnataka. Pavo 32:59–62
Jovani R, Tella JL (2004) Age-related environmental sensitivity and weather mediated nestling mortality in White Storks (*Ciconia ciconia*). Ecography 27:611–618

Kahl MP (1963) Thermoregulation in the Wood Stork, with special reference to the role of the legs. Physiol Zool 36:141–151
Kahl MP (1970) Observations on the breeding of storks in India and Ceylon. J Bombay Nat Hist Soc 67:453–461
Kahl MP (1971a) The courtship of storks. Nat History 80:36–45
Kahl MP (1971b) Spread-wing postures and their possible functions in the Ciconiidae. Auk 88:715–722
Kharitonov SP, Siegel-Causey D (1988) Colony formation in seabirds. Curr Ornithol 5:223–272
Krishnan M (1980) The availability of nesting materials and nesting sites as vital factors in the gregarious breeding of Indian Water-birds. J Bombay Nat Hist Soc 75:1143–1152
Lowther EHN (1949) A bird photographer in India. Oxford University Press, London
Mayfield HF (1961) Nesting success calculated from exposure. Wilson Bull 73:255–261
Mayfield HF (1975) Suggestions for calculating nest success. Wilson Bull 87:456–466
Meganathan T, Urfi AJ (2009) Inter-colony variations in nesting ecology of Painted Stork (*Mycteria leucocephala*) in the Delhi Zoo (North India). Waterbirds 32:352–356
Nagulu V, Rao JVR (1983) Survey of South Indian pelicanries. J Bombay Nat Hist Soc 80:141–143
Naoroji R (1990) Predation by *Aquila* Eagle on nestling storks and herons in Keoladeo Ghana National Park, Bharathpur. J Bombay Nat Hist Soc 87:37–46
Neginhal SG (1977) Discovery of a pelicanry in Karnataka. J Bombay Nat Hist Soc 74:169–170
Neginhal SG (1982) The birds of Ranganathittu. J Bombay Nat Hist Soc 79:581–593
Padmanabhan P, Yom-Tov Y (2000) Breeding season and clutch size of Indian passerines. Ibis 142:75–81
Pai M (2009) The Painted Storks of Veerapuram and Kaggaladu. The vanishing species. http://mohanpaisarticles.blogspot.com/2009/05/vanishing-species-painted-stork.html. Accessed 10 Sept 2010
Pande S (2006) Bhadalwadi tank, a refuge for Painted Storks. Hornbill April–June:11–15
Parasharya BM, Naik RM (1990) Ciconiiform birds breeding in Bhavnagar city, Gujarat. In: Daniel JC, Serrao JC (eds) Conservation in developing countries: problems and prospects. Bombay Natural History Society, Mumbai
Pattanaik C, Prasad SN, Murthy EN, Reddy CS (2008) Conservation of Painted Stork habitats in Andhra Pradesh. Curr Sci 95:1001
Paulraj S, Gunasekaran G (1988) The Vedanthangal water-bird sanctuary: a new breeding ground for pelicans and Painted Storks. J Bombay Nat Hist Soc 85:414–415
Perrins CM, Birkhead T (1983) Avian ecology. Blackie, London
Qiu J (2008) Asian monsoon cycle disrupted by man-made climate change. Nature. Published online 6 Nov 2008. doi:10.1038/news
Rajaram A (1999) An update on the Edayur Bird Sanctuary, Kalpakkam. Newslett Birdwatchers 39:69
Rhenius CE (1907) Pelicans breeding in India. J Bombay Nat Hist Soc 17:806–807
Rodgers JA Jr, Wenner AS, Schwikert ST (1988) The use and function of green nest material by wood storks. Wilson Bull 100:411–423
Sankhala K (1990) Gardens of God. The Waterbird sanctuary at Bharatpur. Vikas Publishing House, New Delhi
Saxena VS (1980) Kokkare Bellur Pelicanry. Cheetal 21:20–24
Shah RV, Desai JH (1975) Growth and development of the Painted Stork, *Ibis leucocephala*, Pennant. II. Post hatching growth pattern and motor development. Pavo 13:88–101
Sharma PK, Raghavaiah PS (2000) Breeding of Painted Stork at Vedurupattu, Nellore district, Andhra Pradesh: co-existence of man and wildlife. Indian Forester 126:1147–1149
Skutch AF (1976) Parent birds and their young. Texas University Press, Austin
Subramanya S, Manu K (1996) Saving the spot-billed Pelican. A successful experiment. Hornbill 2:2–6
Tere A (2008) Nesting of Painted Stork at Palasva village, Kachchh. Flamingo 6:7
Urfi AJ (1993) Heronries in the Delhi region of India. Oriental Bird Club Bull 17:19–21
Urfi AJ (1997) The significance of Delhi Zoo for wild waterbirds, with special reference to the Painted Stork *Mycteria leucocephala*. Forktail 12:87–97

Urfi AJ (1998) A monsoon delivers storks. Nat History 107:32–39
Urfi AJ (2003a) Record of a nesting colony of Painted Stork *Mycteria leucocephala* at Man-Marodi Island in the Gulf of Kutch. J Bombay Nat Hist Soc 100:109–110
Urfi AJ (2003b) The ornithological and conservation significance of village pond heronries. A case study of Seelaj and Traj village ponds around Ahmedabad (Gujarat). Mayura 16:3–9
Urfi AJ (2010) Factors causing nest losses in the Painted Stork (*Mycteria leucocephala*): a review of some Indian studies. J Bombay Nat Hist Soc 107:55–58
Urfi AJ, Meganathan T, Kalam A (2007) Nesting ecology of Painted Stork (*Mycteria leucocephala*) at Sultanpur National Park (Haryana), India. Forktail 23:150–153
Webb-Peploe CG (1945) Notes on a few birds from the south of the Tinnevelly district. J Bombay Nat Hist Soc 45:425–426
Westneat DF, Sherman PW, Morton ML (1990) The ecology and evolution of extra-pair copulation in birds. Curr Ornithol 7:331–369
Wilkinson ME (1961) Pelicanry at Kundakulam, Tirunelveli district. J Bombay Nat Hist Soc 58:514–515
World Meteorological Organization (WMO) (2010) World Weather Information Service. http://worldweather.wmo.int/066/c00224.htm. Accessed 23 Oct 2010

# Chapter 4
# Sexual Size Dimorphism and Mating Patterns

*The males of many birds are larger than the females, and this no doubt is the result of the advantage gained by the larger and stronger males over their rivals during many generations.*

Charles Darwin (1874)

**Abstract** Sexual size dimorphism (SSD) occurs in the vast majority of birds, although explanations for its existence are controversial. This chapter describes the results of a field study undertaken at the Delhi Zoo to establish SSD in the Painted Stork. A novel field study method was used, involving recording of copulating birds and obtaining measurements of tibia, tarsus, bill, body length, and body depth, on each individual in a pair from the images. Males were larger than the females in all of the 100 pairs recorded during the study period. A PCA performed on the available data set of male and female independently showed significant difference's in the first PC scores, which were highly correlated with body length in both sexes. A tendency towards positive assortative mating was observed by plotting the male and female values of different variables. There were no significant differences in the median body lengths of individuals recorded copulating early and those late in the nesting season. However, early male birds were recorded to have significantly larger tarsus, compared to later birds. The second aspect explored in this chapter is spatiotemporal patterns of nesting activity. In the Delhi Zoo, considerable variations were recorded in clutch size and nest initiation dates. Early birds, which colonized pond 2 first, exhibited a higher clutch size. The satellite colonies near the main colonies were always colonized last and had smaller clutch sizes.

## 4.1 Sexual Size Dimorphism and Its Consequences

In most birds and mammals the male is generally larger in size compared to the female. Biologists have long puzzled over how sexual size dimorphism (SSD) evolves and when it comes to selecting a mate then do individuals look for some

A.J. Urfi, *The Painted Stork: Ecology and Conservation*, 
DOI 10.1007/978-1-4419-8468-5_4, 

particular attributes in their partner, possibly a similarity in body sizes? These vexing questions are age old and as humans we are intrinsically curious about what factors influence mate choice in our species. Admittedly, while birds and other animals probably base their judgment on simple rules, in the case of humans the situation is incredibly complex when it comes to selecting a life partner. But ever since Darwin (1874) speculated upon body size differences between the sexes and wrote about them in his book "The decent of man and selection in relation to sex," the interest in the evolution of SSD has not waned and continues to excite evolutionary biologists even today (Andersson 1994; Fairbairn 1997).

In most birds, barring cases of sexually dimorphic plumage and skin coloration pattern, vocalizations and behavioral repertoire, the male is recorded to be larger than the female (Selander 1972). While this difference can be primarily attributed to male–male competition, natural selection acting to reduce competition between the sexes (Slatkin 1984; Shine 1989) and fecundity selection (Schulte-Hostedde and Millar 2000) are alternate evolutionary hypothesis.

Mate preference is cited as a major explanation for SSD. Females may prefer larger males primarily because they are more successful in obtaining territories. In territorial disputes the larger male will always have an edge if his rival contestant is smaller in body size. The second question is: do bigger individuals mate with larger ones and vice versa? In other words, is assortative mating with respect to body size (or some determinant of body size) the norm? Several studies have reported that assortative mating with respect to body size and length of wing, bill, tail, and tarsus is indeed widespread in birds (Coulter 1986; Chardine and Morris 1989; Sandercock 1998; Wagner 1999; Helfenstein et al. 2004) for which a number of underlying mechanisms have been proposed. These include active mate choice, intrasexual competition, mating constraints, or differential mate availability (Sandercock 1998; Helfenstein et al. 2004). However, in some birds, mate selection with respect to body size has also been reported to be random (Bowman 1987).

Although SSD has been studied in several groups of birds (Székely et al. 2007), very few studies have focused on large wading birds (Kushlan 1977; Bildstein 1987). In the family Ciconiidae the information on SSD is scanty, mostly available through morphometric studies on a handful of museum specimens. Among these all the relevant body size parameters do not seem to have been recorded. However, from the available information it appears that in many species of storks the male is larger than the female (Hancock et al. 1992); although in most cases the size differences between the sexes are not easily noticeable. The exception seems to be the Saddlebill in which the female is 10–15% smaller than the male (del Hoyo et al. 1992).

There appear to be few cases of sexual dimorphism among storks. The differences in iris color of Black-necked Stork, brown/red in the case of male and conspicuous bright lemon-yellow in the case of female, is a well known example (Ali and Ripley 1987).

## 4.2 Measuring SSD

At the Delhi Zoo, Painted Stork nests can be seen from a close range, between 43 and 76 m, from the shore and are therefore amenable to a detailed study. Several interesting questions can be asked about these free living storks, such as to what extent do the males and females differ in terms of body sizes? Are any patterns discernible with respect to body sizes when it comes to mate selection? What are the patterns of habitat selection? And so on. In a series of studies undertaken by us at the Delhi Zoo during 2004–2006 some of these issues were addressed.

To establish SSD in a population it often becomes necessary to have physical access to living or dead specimens from which morphometric data can be recorded. In such studies, the sample size has to be reasonably large so that relevant statistical tests can be performed. This is not possible with the Painted Stork; culling large numbers of individuals of a large bird which is on the endangered list for the purpose of a scientific investigation is out of the question and if undertaken, bound to invite sanctions of the strictest kind. Therefore a novel approach in which not even a single bird was touched, killed, or trapped was employed. This approach involved videography from which morphometric data could be easily recorded, was tried out at the zoo. Images of Painted Stork, while they were copulating at the nest, were captured on video (Fig. 4.1). After disengagement following a copulation bout, when the male and female birds were observed standing side by side, one to three still images of each pair were selected for detailed analysis. Video images were converted into metric units by calibrations with museum specimens of Painted Stork. Validation exercises were performed to check for the accuracy of the data generated by the videographic approach, as described in Urfi and Kalam (2006).

Dimensions of the following hard body parts, using the methodology in Bosch (1996) and Wagner (1999) were estimated. All the parameters recorded are shown in Fig. 4.2.

*Bill length*, estimated as the distance from the tip of the upper mandible to the corners of the mouth.

*Tibia length*, estimated as the distance from the joint of the tibia-tarsus till the feathers.

*Tarsus length*, estimated as distance between the tibia-tarsus joint and foot.

On each bird separate measurements were made of the right and left legs. Two additional characters, pertaining to the feathered body parts of the bird, were also estimated.

*Body length*, estimated as the distance from a point corresponding to the base of the neck and the breast to a point roughly corresponding to the origin of the tail.

*Body depth*, estimated as the distance from the highest point on the back to the belly.

During the two nesting seasons 2004 and 2005 about 100 pairs were studied by the videographic method. However, for each character the total number of paired observations was considerably less. This happened because while in most cases one bird in

**Fig. 4.1** Video images of Painted Stork mating in the natural heronries of Delhi Zoo during 2004–2005. (**a**) A pair of storks copulating in the nest. (**b**) Standing side by side after disengagement, permitting identification of the sexes

the pair was captured in a near perfect position, in the case of its mate either the bill was slightly tilted or some feature of its morphology obscured by the surrounding vegetation or the body of its own mate. The resulting data was basically of two types; for any given character there was a set of paired data and another set of unpaired data. This situation provided us with an opportunity to analyze the data in two different

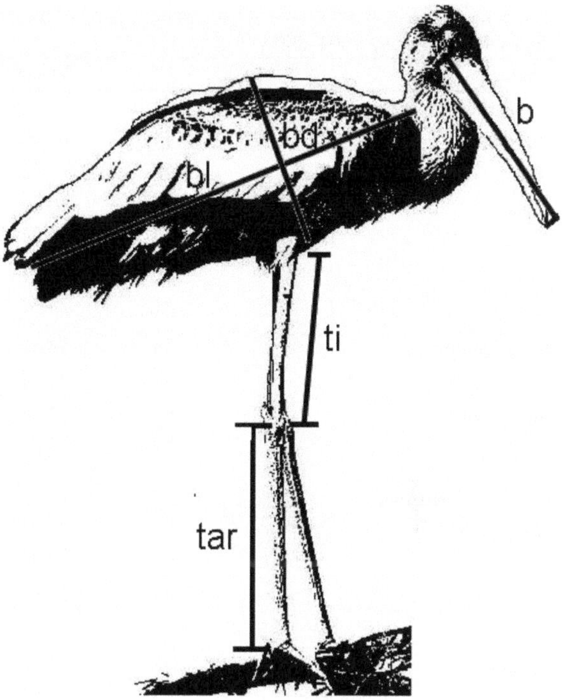


**Fig. 4.2** Figure of Painted Stork to show the body parts that were measured by using the video graphic method. *b* bill length; *bl* body length; *bd* body depth; *ti* tibia; and *tar* tarsus. See text for details

ways. Firstly, we compared the differences in the dimensions of the characters as mean value for each sex and secondly, in the paired sets, in which for the same character there were corresponding values for both the male and female. The former would be helpful in addressing the question, whether in the Painted Stork the male is generally larger than the female and the latter would help us to address the question whether individuals of a particular size copulate with mates of a corresponding size.

In literature several dimorphism indices (DI) are available and their properties have been extensively reviewed (Greenwood 2003). The one used in our study, for any given character, was:

$$\mathrm{DI} = \frac{\text{female mean}}{\text{male mean}} \times 100.$$

## *4.2.1 Exploration of SSD in the Delhi Zoo Population*

Since there was no significant yearly difference in the mean sizes of various body parts, the data for 2004 and 2005 was combined for a detailed statistical analysis. In both the unpaired and paired data sets (Table 4.1), significant differences were observed in the dimensions of the selected body parts of male and female birds. Three points are of

**Table 4.1** Estimated sizes (cm) of various body parts of male and female Painted Stork studied at the Delhi Zoo during 2004–2005

| Character | Sex | *N* | Mean[a] | SD | CV | DI |
|---|---|---|---|---|---|---|
| Unpaired comparisons | | | | | | |
| Bill length | ♂ | 88 | 26.60 | 1.09 | 4.09 | 88.65 |
| | ♀ | 85 | 23.58 | 1.54 | 6.53 | |
| Tibia | ♂ | 68 | 20.90 | 2.06 | 9.86 | 91 |
| | ♀ | 54 | 19.02 | 2.32 | 12.19 | |
| Tarsus | ♂ | 60 | 25.51 | 2.54 | 9.96 | 92.9 |
| | ♀ | 41 | 23.70 | 2.32 | 9.79 | |
| Body length | ♂ | 54 | 52.61 | 7.02 | 13.34 | 90.71 |
| | ♀ | 49 | 47.72 | 6.04 | 12.66 | |
| Body depth | ♂ | 56 | 27.94 | 3.88 | 13.89 | 90.26 |
| | ♀ | 53 | 25.22 | 2.72 | 10.78 | |
| Paired comparisons | | | | | | |
| Bill length | ♂ | 81 | 26.58 | 1.11 | 4.18 | 88.34 |
| | ♀ | | 23.48 | 1.48 | 6.30 | |
| Tibia | ♂ | 48 | 20.81 | 1.98 | 9.51 | 90.1 |
| | ♀ | | 18.75 | 1.99 | 10.61 | |
| Tarsus | ♂ | 36 | 25.96 | 2.29 | 8.82 | 90.64 |
| | ♀ | | 23.53 | 1.96 | 8.33 | |
| Body length | ♂ | 39 | 52.16 | 6.70 | 12.84 | 90.09 |
| | ♀ | | 46.99 | 5.70 | 12.13 | |
| Body depth | ♂ | 45 | 28.38 | 3.92 | 13.81 | 88.97 |
| | ♀ | | 25.25 | 2.67 | 10.57 | |

*Source*: Urfi and Kalam (2006)
*SD* standard deviation; *CV* coefficient of variation (%); and *DI* dimorphism index (%). The paired and unpaired data sets are shown separately
[a]ANOVA $P<0.001$

interest here. Firstly, the variability was higher in the soft body parts compared to the hard skeletal parts. Secondly, the variability in most characters was greater in the case of the female. Lastly, the highest differences were recorded in the dimensions of bill (DI = 88%).

A principal component analysis (PCA) performed on the untransformed data set revealed that the first two components account for 85 and 88% of the variation in the male and female, respectively (Table 4.2). First Principal Component (PC1), which accounts for 66% variation in male and 80% in female, can be interpreted as a size axis in which the maximum correlation is observed in the case of body length. Using PC1 scores as an indication of overall body size, we tested for differences between the male and the female. An ANOVA of the two samples revealed differences ($F=4.46$, df = 1, 46, $P<0.05$) indicating significant SSD (Fig. 4.3). A test of normality revealed that although the PC1 scores of both sexes are normally distributed, the skewness of the female scores is greater than the male scores. When PC1

**Table 4.2** Correlations of the original variables with Principal components I and II of the external characters of male and female Painted Stork.

| | Male (29) | | Female (19) | |
|---|---|---|---|---|
| Variable | PCI | PCII | PCI | PCII |
| Tibia | −0.121 | −0.233 | −0.129 | −0.673 |
| Tarsus | −0.049 | −0.581 | −0.056 | −0.274 |
| Bill | −0.009 | −0.179 | −0.108 | −0.468 |
| Body length | −0.900 | 0.368 | −0.956 | 0.265 |
| Body depth | −0.416 | −0.665 | −0.236 | −0.428 |
| Eigenvalue | 38.663 | 10.878 | 44.843 | 4.327 |
| Proportion of variance | 0.664 | 0.187 | 0.803 | 0.077 |

The sample size for each sex is given in parenthesis
*Source*: Urfi and Kalam (2006)

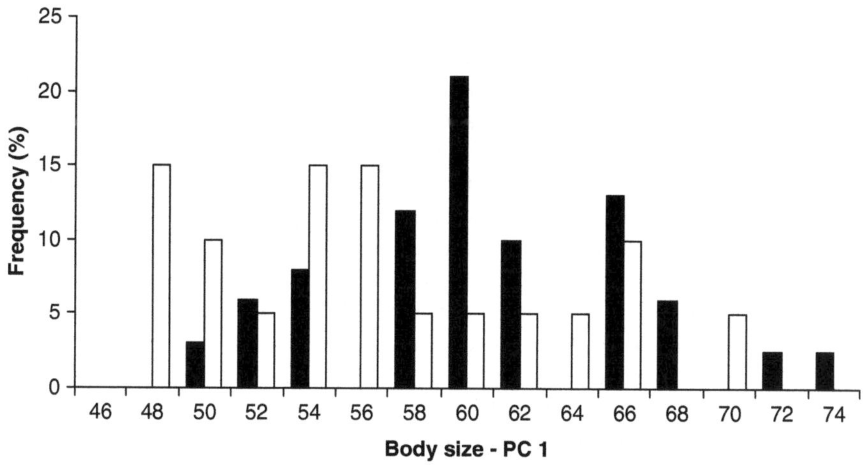


**Fig. 4.3** Frequency distribution of PC1 scores of male (*open squares*) ($n=29$) and female (*filled squares*) ($n=19$) Painted Stork. PC1 scores were obtained from a PCA of five variables of external body parts. *Source*: Urfi and Kalam (2006)

scores were plotted against the different variables, the best linear fit was with body length in both sexes (Fig. 4.4). For the male the regression equation of body length = 0.9001 * PC1 − 3.6303 ($r^2=0.95$, $P<0.001$) and for the female the regression equation of body length = 0.9556 * PC1 − 7.5763 ($r^2=0.99$, $P<0.001$).

The earlier mentioned results clearly show that in the Painted Stork males are larger. While tarsus is a good indicator of body size in many birds (Rising and Somers 1989), differences in other external factors can also be relevant. In this study, PCA shows that body length is a better indicator of overall body size, compared to other characters because of its high correlation with PC1 score in both sexes (Fig. 4.4). The implications of this result are obvious. As our observations on the nesting behavior of Painted Stork in the Delhi Zoo also confirm, most social interactions are intraspecific in nature and their intensity is highest in the beginning of the nesting season. Therefore, it is not difficult to imagine that in territorial fights

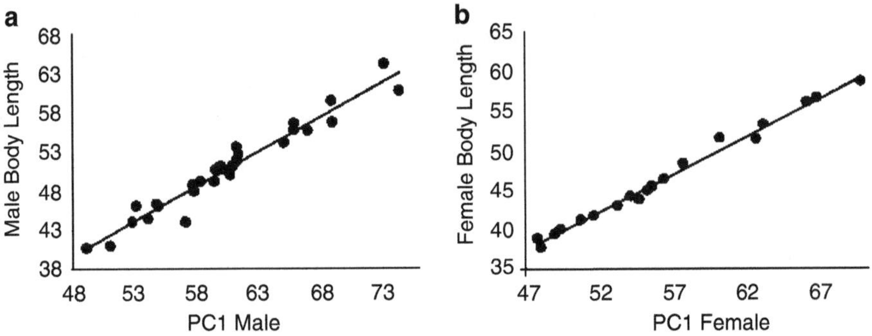


**Fig. 4.4** Male (**a**) and female (**b**) body length (cm) as a function of overall body size as determined by a PCA on five variables of external body parts of male and female Painted Stork. For details see text. *Source*: Urfi and Kalam (2006)

larger individuals will have an advantage over smaller birds. In White Ibis breeding colonies, males are more likely to pirate prey and are less likely to be victims of piracy than females, presumably because they are better able to intimidate others (Frederick 1985). It is important to bear in mind that the 9–10% difference in body length between sexes, in the present study, would translate into a much greater difference in terms of body mass considering the allometric relationship between length and mass in animals. Greater male size would be favored by natural selection although there would be an optimal body size, beyond which greater SSD would be a disadvantage. However, since an increase in body size could also be a function of age this aspect needs to be further investigated. Studies on the relationship between body size of the male and female birds and their fitness components could also be an interesting line of investigation. Interestingly, as we will see later in this chapter, the birds that breed early in the season are larger than the ones which breed later.

The high value of DI in the case of bill length is likely to have important consequences with respect to the foraging abilities of the two sexes. A trophic structure like the bill can be indicative of selection although enormous plasticity has been reported in this structure. Larger bill sizes in the case of males could result in more efficient harvesting of food resources, which in turn could lead to greater fitness (Bildstein 1987). In the case of Painted Stork, as discussed in Chap. 6, the primary demand on the trophic structure may be to get a good grip on a prey which is slippery and strong and capable of slipping off easily once having been caught. Here, a larger bill could be advantageous. A larger bill may be more efficient also in terms of gathering nesting material, such as pulling off the twigs needed for nest construction. In the present study the DI with respect to bill length is 88% and it is quite possible that size dimorphism with respect to bill size could have a role to play in terms of sexual selection. Larger tarsus and bill size may also be helpful in utilization of deeper foraging habitat. It has been suggested that morphological differences can lead to differential foraging which in turn helps in reducing the intersexual competition for food (Selander 1966). To what extent this may also be the case in Painted Stork remains to be seen.

Looking at a single Painted Stork in the field is it possible to tell what sex it belongs to? The only way to sex a monomorphic bird is to capture one alive and take a tissue sample for molecular sexing. Alternatively, using discriminant function analysis (DFA) birds can be sexed provided a detailed analysis has been made of various morphometric parameters beforehand. In both cases physically handling the birds is necessary.

Ratios of various body parts have been useful in establishing SSD or studying its ecological and morphological correlates (Shine et al. 1998; Fessler et al. 2005). Since it is impossible to determine in the field whether a randomly sighted Painted Stork is a male or a female, the possibility of using ratios of the different body parts to sex birds in the field was also explored by us. For this purpose 10 ratios from the data available were calculated separately for the male and the female. In all cases the differences between the ratios for each set of characters was not statistically significant (*t*-test, n.s.), suggesting that field sexing is not possible with the available data set (Urfi and Kalam 2006).

## 4.3 Assortative Mating

Large sized birds would be expected to mate with large individuals and vice versa. To study mate size selection pattern in the Painted Stork we plotted the values of four different body variables of the male against those of the female, using paired data sets. The details are given in Fig. 4.5 and the linear regression equations for each case are as follows:

Female body length $=16.0+0.595$ male body length, $r^2=0.49$, $P<0.001$
Female bill length $=0.73+0.856$ male bill length, $r^2=0.41$, $P<0.001$
Female tibia $=8.03+0.515$ male tibia, $r^2=0.26$, $P<0.001$
Female tarsus $=15.3+0.315$ male tarsus, $r^2=0.13$, $P<0.05$

These results are indicative of positive assortative mating.

Several factors have been attributed to the evolution of assortative mating namely, active mate choice, intrasexual competition, and mating constraints/differential mate availability (Coulter 1986; Sandercock 1998; Wagner 1999) and the matter needs to be explored in detail in the case of storks. Interestingly, while there is a clear-cut hierarchy in the sizes of the males, there is considerable variation in female sizes. The variability is observed in other characters as well (Fig. 4.5). This variability could be due to biased sex ratios in the Painted Stork population, with there being more females than males. While more studies are required on mating patterns an important point to be borne in mind, as far as this study is concerned, is that the data obtained from videography was essentially morphometric recordings of birds sexed during copulation. It was assumed that the pair engaged in copulation was a pair associated with a particular nest but the possibility of two individuals engaged in an extra-pair copulation also cannot be ruled out.

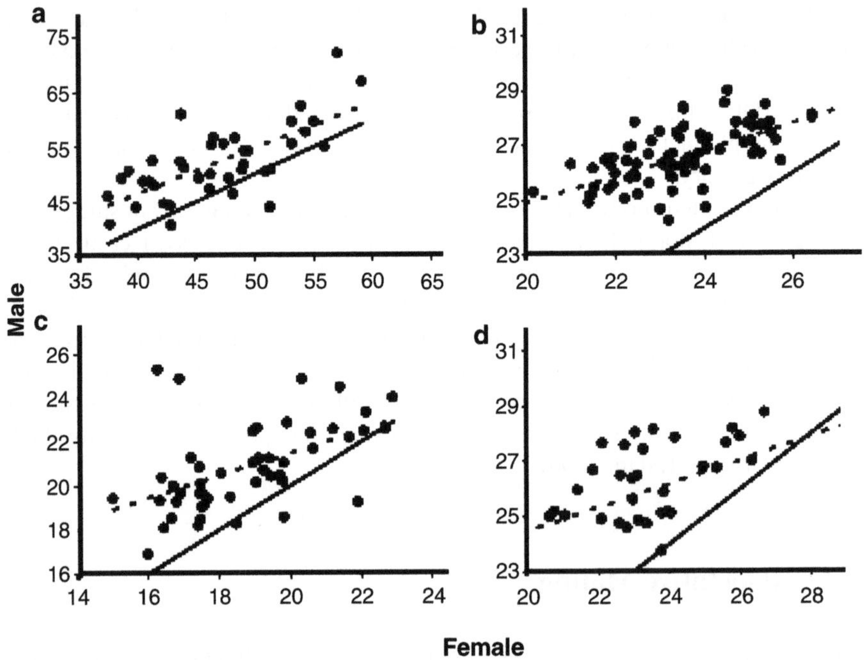


**Fig. 4.5** Relation of lengths (in cm) of: (**a**) body length, (**b**) bill, (**c**) tibia, and (**d**) tarsus of Painted Stork recorded on video during copulation at Delhi Zoo in 2004–2005. The *dotted lines* shows the trend line (significant at $P<0.001$ in all cases) and the *solid lines* shows the condition $x=y$ in all graphs. Note that the abscissa and ordinate scales are different in each graph. *Source*: Urfi and Kalam (2006)

## 4.4 Patterns of Habitat Selection

To a casual visitor seeing Painted Stork nesting on trees in the premises of the Delhi Zoo, it will not be immediately obvious that there are some patterns behind how the different trees were occupied. Our studies have led us to some degree of understanding about the underlying rules which seem to operate when it comes to the sequence of events in the breeding phase. The Delhi Zoo is a system that we are slowly beginning to understand and it can become a model system for detailed studies later on (Urfi 2010).

The data generated by the videographic approach used for SSD studies could also be employed for other purposes too. This combined with some other field studies subsequently became extremely helpful in addressing a whole bunch of other questions. Since the date of each copulation was recorded, it was possible to group individuals into early and late breeders. Using the 2004 data only, since data on all variables was available for a longer period during this year, we classified early nesters as those individuals which were recorded copulating before 1 September. The rest of the data was grouped under late nesters and the two sets of data were used for studying the differences between early and late nesters. Interestingly, with respect to the variable body length, no significant differences were observed in either the male or female. However, using tarsus length as a criteria we found that median tarsus length of males among the early nesters was significantly greater (Mann Whitney test $P=0.001$) than the later birds. In the case of the females the differences between the tarsus lengths of early and late birds was not significantly different. Whether the late birds were those individuals which started early and either failed in nesting and were starting another brood or attempting extra-pair copulations was difficult to ascertain since the zoo birds were not individually marked.

The studies detailed previously open up a number of other questions. For instance, at the time of commencement of nesting, do Painted Stork choose only some specific trees as substrate? What are the differences between the trees selected and those ignored? What are the sequences of events by which the zoo colonies are utilized? In a separate study (Meganathan and Urfi 2009) we attempted to study these aspects of Painted Stork breeding in the Delhi Zoo.

Table 4.3 shows the two main ponds (1 and 2) of the zoo that have two islands each on which a number of mesquite trees have been planted. The merged canopies of trees in the first pond are large as compared to those in the second pond. In fact, in one case, the total canopy area is 1,445 $m^2$. In this pond the trees which constitute the substrate for nesting colonies are taller and correspondingly the average height of nests is also greater. However, the nest density of pond 2 colonies is higher than that of pond 1.

Besides the islands in the two ponds, Painted Stork also builds nests on neighboring trees in the vicinity of ponds. For reasons of convenience we have given each of these smaller colonies a name in relation to the nearest main colony. For instance, the colony nearest to the main colony 3 is called as satellite 3 or S3. The reason for naming it so is simply due to the fact that it is in the immediate vicinity of colony 3 and seems like an extension of it. Simplistic or probably inaccurate as this assumption might be it brings up a fundamental question that has been plaguing colony researchers for a long time, which is, how large a gap is necessary between nesting birds before one colony becomes two (Coulson and Dixon 1979 in Jovani and Tella 2007)? Is this only a methodological problem? The existence of colonies is considered to be beyond question as we have already discussed in Chap. 2. However, according to Jovani and Tella (2007) the distribution of nests and how it is organized through spatial scales can provide important insights about colony structure.

In the Delhi Zoo, in each of the years of study we found a characteristic colonization pattern. In August, at the time when Painted Stork start congregating at the zoo, the first arrivals settle on Colonies 3 and 4 in Pond 2. As previously mentioned, these early birds were bigger and presumably in good condition. This would mean that these colonies were highly valued and probably offered the best habitat.

## 4.5 Variations in Clutch Size and Nest Initiation Dates

As we subsequently discovered, some key fitness parameters among the early nests on prime colonies was significantly different from those in other colonies as well as late nesting birds.

**Table 4.3** The characteristics of the main and satellite colonies of Painted Stork in the Delhi Zoo during 2005–2006

| Colonies | Tree species | Colony area ($m^2$)[a] | Tree height (m)[b] | Nest height (m)[b] | Nest density[c] | Clutch size[d] | |
|---|---|---|---|---|---|---|---|
| | | | | | | 2005 | 2006 |
| 1 Island (Pond I) | Mesquite (17) | 572.78 | 8.43 | 4.16±0.37 (2) | 0.45±0.18 (2) | 1.75±0.21 (12) | 1.80±0.20 (10) |
| 2 Island (Pond I) | Mesquite (28) Peepul (1) | 1445.3 | 7.96 | 6.05±0.22 (4) | 1.25±0.16 (4) | 2.42±0.15 (19) | 2.00±0.17 (20) |
| 3 Island (Pond II) | Mesquite (14) | 491.07 | 5.18 | 3.17±0.12 (8) | 2.33±0.13 (3) | 3.04±0.12 (48) | 2.76±0.18 (25) |
| 4 Island (Pond II) | Mesquite (23) | 638.19 | 6.87 | 3.70±0.09 (7) | 1.67±0.13 (4) | 2.70±0.13 (34) | 2.24±0.16 (25) |
| S1 On shore (Pond I) | Mesquite (2) | 210.29 | 8.5 | 7.40±0.13 (5) | 0 | – | – |
| S3 On shore (Pond II) | Mesquite (3) | 346.5 | 8.3 | 7.71 (1) | 1.00±0.24 (7) | – | – |
| S4 On shore (Pond II) | Mesquite (2) | 186.34 | 9.4 | 7.12±0.36 (7) | 0.71±0.18 | – | – |
| Satellite colonies (combined) | – | – | – | – | – | 1.52±0.15 (17) | 1.41±0.19 (12) |

Sample size for some parameters given in parenthesis. Details about methodology provided below. See footnote for details

Methodology: Field studies were conducted, 5 days per week, from 06:30 to 18:00, from August 2005 through December 2006. Photographs were taken of all colonies from approximately 20–65 m. Nests in the photographs were serially numbered and 151 and 101 nests, during 2005 and 2006, respectively were randomly selected for a detailed study. Relocation of the preselected nests on each visit was accomplished by referring to the photographs

*Source*: Meganathan and Urfi (2009)

Nesting habitat characteristics were recorded as follows:

[a]*Colony area*: For each island, the dimensions of the two perpendicular axes were estimated using the GIS software Google Earth, after due calibrations on the ground. Assuming that the colonies were circular, the mean of the two axes was divided by two to obtain radius dimensions from which area was calculated

[b]*Nest height*: Since in photographs of colonies there were inevitably Painted Storks standing in a position in which their bill could be easily photographed in full length. Bill length (approximately 28 cm) was used as a scale to estimate tree and nest height

[c]*Nest density*: It was estimated as number of nests inside an imaginary circle, of radius 3 bill length, around the focal nest. A preliminary exercise was undertaken in which marked objects were placed at varying distances which were then estimated in relation to a cardboard model of a life size Painted Stork bill. This exercise helped us to ascertain the accuracy of our method and also to initially practice estimation of distances in three dimensions with the Painted Stork bill as a scale

[d]*Clutch size*: It was taken as the number of eggs in a nest after the last egg had been laid, and was estimated by direct observation using binoculars or a telescope. If the focal nest was too high the investigator climbed a nearby tree to obtain a better view of nest contents

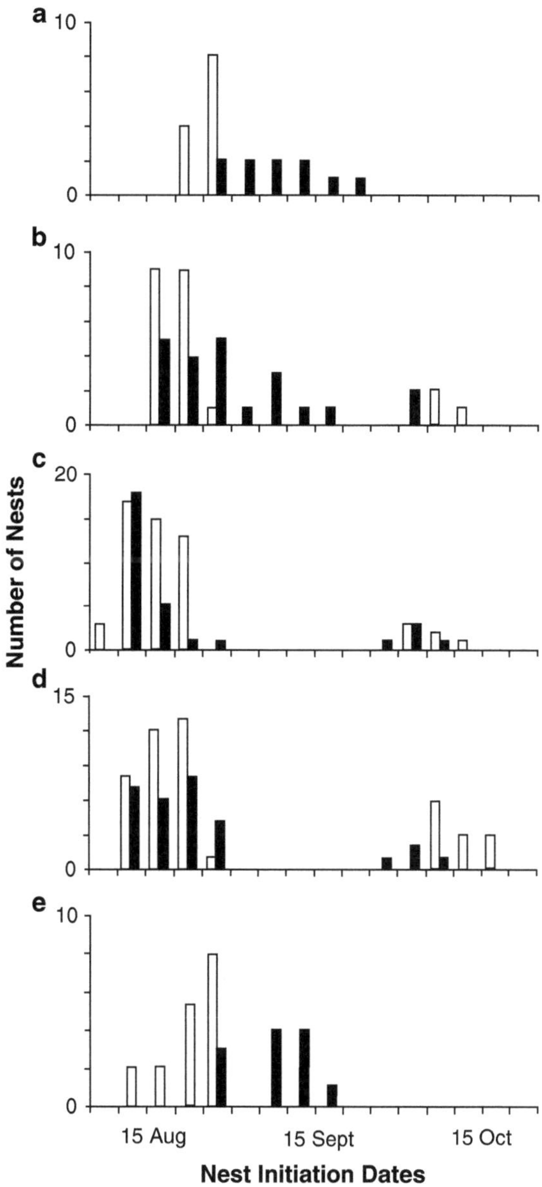


**Fig. 4.6** Variations in nest initiation date of Painted Stork in the Delhi Zoo colonies during 2005 (*open squares*) and 2006 (*filled squares*). (**a**) Colony 1, (**b**) colony 2 (both in Pond 1), (**c**) colony 3, (**d**) colony 4 (both in Pond 2), and (**e**) combined satellite Colonies. *Bars* represent number of nests initiated in 5-day intervals within each month. Note the scale differences on the vertical axes. All the three satellite colonies have been grouped together. The nest initiation date was determined on the basis of the first egg laid. *Source*: Meganathan and Urfi (2009)

Nest initiation dates differed significantly across colonies (ANOVA $P<0.001$ in both 2005 and 2006). Colonies 3 and 4 were always the first ones where nests were initiated (Fig. 4.6). In contrast among the satellite colonies the nests were initiated very late ($t$-test, $P<0.001$ in both years). In some colonies a second bout of nesting activity was also recorded (Fig. 4.6).

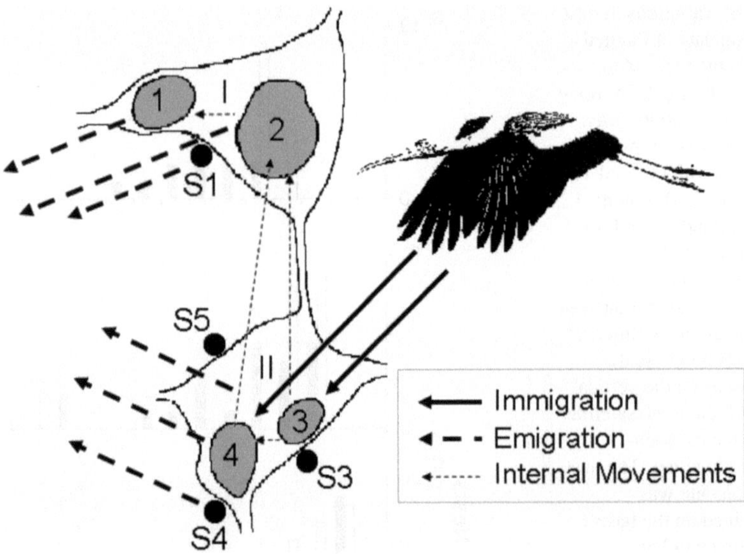


**Fig. 4.7** A diagram to show the sequence of occupation of different colonies of the Delhi Zoo during 2005–2006. *Source*: Meganathan and Urfi (2009)

The mean clutch size for the whole zoo was recorded to be 2.43±0.07 ($n$=151) and 2.09±0.09 ($n$=101) in 2005 and 2006, respectively, differed significantly between the 2 years (t218=2.86, $P$=0.005). There were marked yearly variations in clutch size too ($t$-test $P<0.005$). Marked variations were also observed in the mean clutch size across colonies (ANOVA $P<0.005$). The mean clutch size of Pond II colonies was higher than Pond I colonies ($t$-test $P<0.001$ in 2005 and $P<0.05$ in 2006). Also, the clutch size of island colonies was greater than satellite colonies (Table 4.3). The mean clutch size of early nests was significantly higher (range 2.42–3.42) than those of late nests (1.28–2.00).

The spatial variations in clutch size, as observed in this study, have also been observed in other species of colonially nesting waders and the phenomenon is ecologically interesting (Delord et al. 2003; Frederick and Spalding 1994). Since among the seven colonies of the zoo, colony 3 is occupied first and exhibits the highest values of clutch size, it would appear to be the best nesting patch. Interestingly, nest density and clutch size are correlated. To what extent this could be related to coloniality in Painted Stork merits investigation. In future, modeling exercises (Dinsmore et al. 2002), involving all the factors and their influence on nesting success, will be extremely useful in terms of understanding habitat selection by Painted Stork (Rendón et al. 2001).

Based on the studies undertaken to explore SSD and mating patterns of Painted Stork at the Delhi Zoo (Urfi and Kalam 2006) and our observations on differences across the different heronries (Meganathan and Urfi 2009) it is possible to visualize the pattern of colonization and the sequence in which different types of nesting substrates are inhabited. A diagrammatic representation of the same is given in Fig. 4.7. At the commencement of the breeding season in late August, the islands in

Pond 2 are among the first to be colonized. Apparently, most birds make a beeline for colonies 3 and 4 and the spillover first goes to islands in Pond 1 (colonies 1 and 2) and then to the satellite colonies. Finally, after all the habitats are occupied those birds which were unable to secure breeding territories emigrate out of the zoo and only those birds which are nesting remain along with a few extra ones.

In the end, several questions remain and future studies should attempt to address these at the Delhi Zoo and other sites. First, why do Painted Stork build nests in certain trees and not in others? The whole of the Delhi Zoo campus is a well wooded area and there are a number of trees of different varieties. In the end Painted Stork choose only a few, surrounded by water and ignore those on the mainland. While obviously this may be influenced by predation pressure from the ground, as discussed in Chap. 3, other factors may be important too. These could be historical factors and past associations of individuals to a particular site, as has been suggested by various authors (Frederick and Meyer 2008). At Keoladeo, where Painted Stork utilizes only some clumps of trees growing in the marsh only in certain blocks of the park, leaving other areas untouched, strongly suggests that this species exhibits "active coloniality" and is not colonial due to the limitation of nesting sites.

At the Delhi Zoo since colonies 3 and 4 of Pond number 2 are colonized first, they would be the best habitat apparently. However, we still do not know what characterizes habitat quality. Is it the shape of the canopy of the merged trees? For the benefits of coloniality to be felt, is there an optimal colony size or optimal nest density? How significant is the impact of aerial predators? We discovered that the numbers of crows and kites, two chief aerial predators at the Delhi Zoo, were slightly higher in the vicinity of Pond 1 and less in Pond 2 (Meganathan and Urfi 2009). Is this likely to be a factor? Clearly this is an area which needs more investigation.

## References


Ali S, Ripley SD (1987) Handbook of the birds of India and Pakistan together with those of Bangladesh, Nepal. Oxford University Press, New Delhi

Andersson M (1994) Sexual selection. Princeton University Press, Princeton

Bildstein KL (1987) Energetic consequences of sexual size dimorphism in White Ibises (*Eudocimus albus*). Auk 104:771–775

Bosch M (1996) Sexual size dimorphism and determination of sex in yellow-legged gulls. J Field Ornithol 67:534–541

Bowman R (1987) Size dimorphism in mated pairs of American Kestrels. Wilson Bull 99:465–467

Chardine JW, Morris JW (1989) Sexual size dimorphism and assortative mating in the Brown Noddy. Condor 91:868–874

Coulson JC, Dixon F (1979) Colonial breeding in seabirds. In: Larwood G, Rosen BR (eds) Biology and systematics of colonial organisms. Academic, London

Coulter MC (1986) Assortative mating and sexual dimorphism in the Common Tern. Wilson Bull 98:93–100

Darwin C (1874) The decent of man and selection in relation to sex, 2nd edn. John Murray, London

del Hoyo J, Elliott A, Sargatal J (eds) (1992) Handbook of the birds of the world, vol 1. Lynx Edicions, Barcelona

Delord K, Kayser Y, Barbraud C, Hafner H (2003) Variability in breeding parameters of the Squacco Heron *Ardeola ralloides*. Bird Study 50:300–305

Dinsmore SJ, White GC, Knopf FL (2002) Advanced techniques for modeling avian nest survival. Ecology 83:3476–3488

Fairbairn DJ (1997) Allometry for sexual size dimorphism: pattern and process in the coevolution of body size in males and females. Annu Rev Ecol Syst 28:659–687

Fessler DMT, Haley KJ, Lal RS (2005) Sexual dimorphism in foot length proportionate to stature. Ann Hum Biol 32:44–59

Frederick P (1985) Intraspecific food piracy in White Ibises. J Field Ornithol 56:413–414

Frederick P, Meyer KD (2008) Longevity and size of wood stork (*Mycteria americana*) colonies in Florida as guides for an effective monitoring strategy in the southeastern United States. Waterbirds 31:12–18

Frederick PC, Spalding MG (1994) Factors affecting reproductive success of wading birds (Ciconiiformes) in the Everglades ecosystem. In: Davis S, Ogden J (eds) Everglades: the ecosystem and its restoration. St. Lucie Press, Florida

Greenwood JG (2003) Measuring sexual size dimorphism in birds. Ibis 145:124–126

Hancock JA, Kushlan JA, Kahl MP (1992) Storks, ibises and spoonbills of the world. Academic, London

Helfenstein F, Danchin E, Wagner RH (2004) Assortative mating and sexual size dimorphism in Black-legged Kittiwakes. Waterbirds 27:350–354

Jovani R, Tella JL (2007) Fractal bird nest distribution produces scale-free colony sizes. Proc Biol Sci 274:2465–2469. doi:10.1098/rspb.2007.0527

Kushlan JA (1977) Sexual dimorphism in the White Ibis. Wilson Bull 89:92–98

Meganathan T, Urfi AJ (2009) Inter-colony variations in nesting ecology of Painted Stork (*Mycteria leucocephala*) in the Delhi Zoo (North India). Waterbirds 32:352–356

Rendón MA, Garrido A, Ramírez JM, Rendón-Martos M, Amat JA (2001) Despotic establishment of breeding colonies of Greater Flamingos *Phoenicopterus ruber*, in southern Spain. Behav Ecol Sociobiol 50:55–60

Rising JD, Somers KM (1989) The measurement of overall body size in birds. Auk 106:666–674

Sandercock BK (1998) Assortative mating and sexual size dimorphism in Western and Semipalmated Sandpipers. Auk 115:786–791

Schulte-Hostedde AI, Millar JS (2000) Measuring sexual size dimorphism in the Yellow-pine Chipmunk (*Tamias amoenus*). Can J Zool 78:728–733

Selander RK (1966) Sexual dimorphism and differential niche utilization in birds. Condor 68:113–151

Selander RK (1972) Sexual selection and dimorphism in birds. In: Campbell B (ed) Sexual selection and the descent of man 1871–1971. Aldine, Chicago

Shine R (1989) Ecological causes for the evolution of sexual dimorphism: a review of the evidence. Q Rev Biol 64:419–461

Shine R, Keogh S, Doughty P, Giragossyan H (1998) Costs of reproduction and the evolution of sexual dimorphism in a "flying lizard" *Draco melanopogon* (Agamidae). J Zool Lond 246:203–213

Slatkin M (1984) Ecological causes of sexual dimorphism. Evolution 38:622–630

Székely T, Lislevand T, Figuerola J (2007) Sexual size dimorphism in birds. In: Fairbairn D, Blanckenhorn WU, Székely T (eds) Sex, size and gender roles. Oxford University Press, Oxford

Urfi AJ (2010) Using heronry birds to monitor urbanization impacts: a case study of Painted Stork *Mycteria leucocephala* nesting in the Delhi Zoo, India. Ambio 39:190–193

Urfi AJ, Kalam A (2006) Sexual size dimorphism and mating pattern in the Painted Stork (*Mycteria leucocephala*). Waterbirds 29:489–496

Wagner RH (1999) Sexual size dimorphism and assortative mating in Razorbills (*Alca torda*). Auk 116:542–544

# Chapter 5
# Growth and Development

*The young of the Painted Stork may be a pretty bird with his pencilled plumage, but he is a noisy brute and seems to spend most of his time trying to let everyone for miles around know how hungry he is.*

C.E. Rhenius (1907)

**Abstract** This chapter details the posthatching growth and development patterns of Painted Stork. The nestlings can be characterized as semi-altricial-1, a type of development pattern which is characterized by an overall high growth rate and an uneven pattern of growth of certain body parts. In the Painted Stork while culmen, middle toe, and tarsus grow at a steady pace, the largest primary grows slowly at first but its size shoots up after about 20 days posthatching. An illustrated account for aging nestlings in the field by observing their behavior and morphology is provided. The broad outlines of moulting patterns, particularly the changes associated with feathering patterns of the head at different ages, is also briefly discussed.

## 5.1 Development Patterns in Birds

Embryonic differentiation and nestling growth are critical stages during the course of development of an individual bird and have therefore always interested ornithologists. It was recognized long ago that the basic patterns of development of hatchlings vary among different birds. In some, like the fowl or duck, the young ones – chicken or ducklings which emerge out of the egg – are miniature versions of their adult forms, fully capable of fending for themselves. On the other hand in some other birds, say the sparrow, the hatchling is a weak, helpless individual, with closed eyes. Its body movements are feeble and uncoordinated and so it has to be taken care of by the parents till it is old enough to fend for itself.

Oken in 1816 (in Starck and Ricklefs 1998) first used the terms *nidifugous* (*Nidi* from Latin *nidus* meaning "nest" and *fugous* from *fugere* meaning "flee") and

A.J. Urfi, *The Painted Stork: Ecology and Conservation*,
DOI 10.1007/978-1-4419-8468-5_5, © Springer Science+Business Media, LLC 2011

*nidicolous* (*colous* from Latin *colere*, meaning "inhabit") to denote offspring who were nest fleers and nest squatters, respectively. The original terms used were *Nestflüchter* and *Nesthocker* but later the terms *altricial* and *precocial* were introduced in ornithological works, as synonyms of *nidicolous* and *nidifugous*. In modern ornithological texts while *altricial* and *precocial* primarily refer to the development stage of the chick, *nidicolous* and *nidifugous* refer to their nest attendance.

The criteria for characterizing the young of any species as being *altricial* or *precocial* are multiple so in reality there is an "altricial-precocial spectrum" where that species (its young) can be placed. Nice (1962) attempted to clarify the picture by describing in detail the different diagnostic features of hatchlings for a variety of birds so that they can be placed in discrete developmental categories. The diagnostic features in this scheme were as follows:

*Plumage of hatchlings*: None or with down or contour feathers
*Eyes*: Open or closed
*Nest attendance*: Stay in the nest, stay in the nest area or leave the nest area
*Type of parental care*: None, brooding, food showing, or parental feeding

Where does the Painted Stork, or for that matter storks in general, stand in this scheme? According to Nice's criteria, Painted Stork nestlings would be semi-altricial-1 because they are covered with down at the time of hatching, have open eyes, stay in the nest till they are quite grown, and during the period are fed entirely by their parents (Shah and Desai 1975b). The Wood Stork is placed in the same category and probably other storks too.

## 5.2 Eggs and Clutch-Size Variations

The fertilization of the egg occurs in the infundibulum of the female and the development of the embryo inside the egg starts immediately. At the Delhi Zoo colony 34 eggs of Painted Stork, marked on the day they were laid, were studied at 3-day intervals by removing them from the nest (Shah and Desai 1975a). On the first day, the blastoderm was recorded as a disk about 4–5 mm in diameter. By day 6, the length of the embryo was recorded to be 25 mm and its weight was 0.9 g. By this stage the eyes, limb buds, and vitelline veins were well developed and the jaws were in the process of being formed.

Painted Stork eggs, when laid, are dull sullied white, sometimes sparsely spotted and streaked with brown (Ali and Ripley 1987). They are generally oval in shape, although elliptical ones are also encountered (Desai et al. 1977). Freshly laid eggs look chalky white but soon assume a dull color, with stain and dirt, after a few days in the nest. A total of 60 eggs at the Delhi Zoo colony were measured and weighed during 1967, 1968, and 1970. Their sizes ranged from 66 to 75 mm in length (long axis) and from 41.5 to 48 mm at the maximum width (short axis). The mean sizes recorded were 69.58 (±2.05) × 43.92 (±1.49) mm which is similar to the values reported by Baker (1932), i.e., 69.5 × 49.0 mm ($n$ = 50).

Studies at the Delhi Zoo revealed that egg weights in the Painted Stork were quite variable. Overall, egg weight, at the time of laying, averaged $74.0 \pm 0.64$ g ($n=60$), range: 66–85 g. As a proportion of the body weight (taken as 3.18 kg for the Painted Stork), the egg weight is approximately 2.45%. The range expected in the stork family is 2–4% (Lack 1968). The shell of the Painted Stork egg is recorded to be thick, tough, and comparatively heavier than a hen's egg. The different components of the egg by weight were estimated as, shell = 9.15%, yolk = 13.8%, and albumin = 77%. These proportions compare well with the data recorded for altricial birds (Desai et al. 1977). At the Delhi Zoo colony, eggs in a clutch were laid 2–4 days apart, mostly in the early hours of the day between 7 and 10 A.M., although sometimes the time elapsed between two successive eggs was even more. Thus, egg laying and subsequently hatching are asynchronous in the case of Painted Stork.

Considerable variations have been recorded in clutch sizes of Painted Stork from different locations. Baker (1932) mentions 3–5 (mostly 6) while Ali and Ripley (1987) mention 2–5 (mostly commonly 3–4). At the Delhi Zoo Desai et al. (1977) recorded the range as 1–4 (most commonly 2–3), the mean lying between 2.2 and 2.8. Meganathan and Urfi (2009) found considerable variations in clutch size both spatially and temporally. Birds which initiated nests early had higher clutch sizes compared to late nesters. Nests on main colonies had higher clutches compared to those in secondary (or satellite) colonies. Clutch size is an important fitness parameter in conservation assessment. However, it remains to be established if it, or other parameters such as egg weight or volume, has changed in relation to deteriorating quality of the habitat in and around the Delhi Zoo due to urbanization and pollution (Urfi 2010).

Incubation starts as soon as the first egg is laid and subsequently both sexes share the task of incubation. According to Desai et al. (1977), the female shoulders the major task of incubation. The mean incubation period for Painted Stork at the Delhi Zoo has been estimated to be 30 days ($n=37$, range 28–32 days).

## 5.3 Growth and Development Patterns in the Painted Stork

The growth and development patterns of several species of storks have been studied in considerable detail (Kahl 1962, 1966). The same was studied in the Painted Stork by Desai and coworkers at the Delhi Zoo during 1966–1971 (Shah and Desai 1975b).

### *5.3.1 Growth Rate*

The growth patterns of altricial and precocial chicks are different, because both strategies have evolved in relation to different environmental constraints. In the case of latter, growth rate is slower, even though the starting value of body mass is higher. Also posthatching growth in precocial birds is extended over a longer period of time, compared to altricial birds.

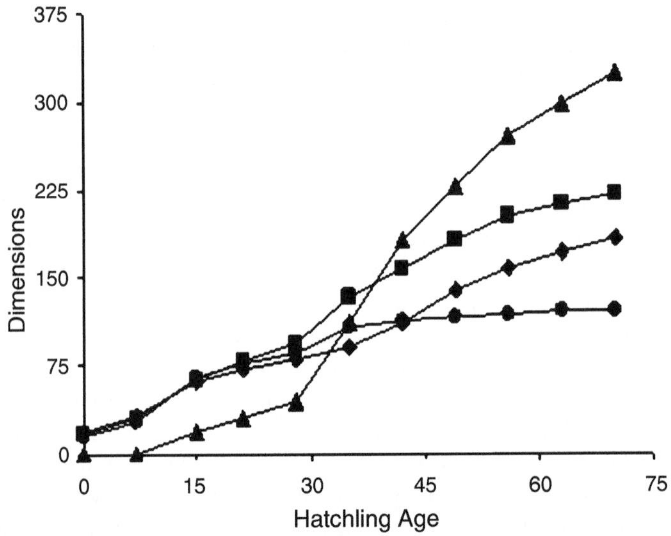


**Fig. 5.1** Growth (mm) of select body parts of Painted Stork hatchlings at different ages (days posthatching). Symbols used *filled triangles* longest primary, *filled squares* tarsus, *filled diamonds* culmen, *filled circles* middle toe. Note the growth rate of the longest primary shoots up after 30 days posthatching. Redrawn from Shah and Desai (1975b)

Posthatching growth patterns in birds are typically described by fitting empirical data of postnatal mass increase to sigmoid growth functions. For estimating the crucial parameter, the rate constant of the equation (*K*), the logistic growth function is commonly used although other functions such as the Gompertz function and the Bertalanffy function have also been employed. Staark and Ricklefs (1998) estimated the growth parameters for a number of birds, including five species of storks. For the Wood Stork with an initial body mass = 2,500 g, the value of $K_G$ (Gompertz function) = 0.085, and $K_L$ (adjusted to Logistic Growth Function) = 0.125. Compared with a precocial, large-sized North American wader, the Sandhill Crane whose chick at the time of hatching weighs about 5,118 g, slightly more than double the weight of Wood Stork nestling, $K_G = 0.032$ and $K_L = 0.047$.

Typically, in the sigmoid growth curves of birds, there is an early phase of rising slope and a later phase of declining slope. The segment corresponding to the rising slope is designated as the "self-accelerating" phase and the other segment is designated as the "self-inhibiting" phase. In the case of the Painted Stork, the self-accelerating phase lasts up till the first 2–3 weeks after hatching and after about 21 DPH the self-inhibiting phase begins (Shah and Desai 1975b).

One characteristic of posthatching growth in semi-altricial birds like the Painted Stork is the uneven pattern of growth rates of different body parts. For instance, while the culmen, middle toe, and tarsus grow at a steady pace, the growth of the longest primary is initially very slow but shoots up after about 30 DPH (Fig. 5.1). Since growth is an energy consuming process, the priority in development is the growth of essential organs first and postponement of growth of those organs which are of use at a later stage.

## 5.4 Posthatching Development

There is a need to have a clear idea about how Painted Stork nestlings look like at different stages of their growth. From a perusal of literature, it emerges that often the age of nestlings of Painted Stork has been defined in loose terms, using such vague expressions as "freshly hatched chicks" or "very young ones" or "considerably older chicks," etc. Such terms are not very useful if one wants to pinpoint exactly when the nesting period started at any given site. However, if the nestling age was clearly mentioned then it is possible to pinpoint the initiation of the nesting period with reasonable accuracy. Such information can be extremely useful in ecological and conservation studies.

In a detailed study undertaken at the Delhi Zoo colonies, Shah and Desai (1975b) elaborated upon the various posthatching growth stages of the Painted Stork. The main points of their study are outlined. Keeping in view the need to have some sort of a key to aging nestlings reasonably accurately an illustrated synopsis (Table 5.1) is also provided here. To keep matters simple, the various growth stages in the first year have been divided into seven stages, each distinct in terms of plumage and behavior development characteristics, starting with the just-hatched nestling (upto 24 h).

1. Just hatched (Fig. 5.2a)
   *Age*: upto 24 h
   *Weight*: 55 g
   *Morphology*: The entire body is sparsely covered with light gray down feathers (prosoptiles). It is thickest on the dorsal surface of the wings but the ventral surface is almost naked. Eyes are open and black in color. The tip of the bill is yellow.
   *Behavior*: The hatchling appears feeble and weak and its movements are uncoordinated. It is neither able to stand on its feet nor able to lift its disproportionately large head properly. Most of its time is spent sleeping in the nest. Occasionally, it emits loud squeaks. The hatchling is not able to pick up food in the nest.
2. Very young nestling (Fig. 5.2b)
   *Age*: 1–15 DPH (1–2 weeks)
   *Weight*: 144–526 g
   *Morphology*: From 6 to 10 DPH the coat of down on the dorsal side starts becoming dense. By 11–15 DPH, prosoptiles are replaced with dense wooly white second generation of down (mesoptiles).
   *Behavior*: The nestlings squeak when any object approaches them. The squeaks are much louder now. From 6 to 10 DPH, they are capable of moving their head and neck at will. Jabbing movements of bill seen while picking up fish from nest. By this time nestlings have acquired the capacity of short movements in nest on tarsi. As days pass, their muscle tone develops and they start exhibiting coordinated actions like preening and cleaning of wings. A tendency to bask in the sun with wings spread out is seen. Also, up–down display movements can also be sometimes observed. By now the nestlings are able to take the fish, about 7–9 cm long, which the parents regurgitate in the nest.
3. Snowball or cotton ball stage (Fig. 5.2c)

**Table 5.1** Synopsis of development of nestlings and their diagnostic morphological features discussed in the text

| Stage | Age | Name/term | Plumage and morphology | Behavior | Figures |
|---|---|---|---|---|---|
| 1 | Upto 24 h posthatching | Just hatched | Light gray down. Yellow tipped beak, eyes open | Feeble, weak, movements uncoordinated | 5.2a |
| 2 | 1–15 DPH | Very young nestling | Coat of down. Bill black | Squeaking noises. Some body movements made including up–down displays. Able to stand on tarsi | 5.2b |
| 3 | 16–30 DPH | Snowball/cotton ball | Body covered with soft, wooly dense down. Black bill and head | Muscle tone improved coordinated movements. Still stands only on tarsi | 5.2c |
| 4 | 31–45 DPH | "Turkey" | About half of adult size. Big patches of black visible on head and neck. Bill still black but size of yellow tip has increased | Very noisy and now able to stand on their feet. Very active and make threat displays at neighbors | 5.2d |
| 5 | 46–60 DPH | Gray young-1 | Head and neck smokey gray. Beak which was so far black now turning lighter in color | Wing flapping and bouncing off in the air while at nest. Local flights undertaken in the colony. Still noisy but the bass quality of voice appearing | 5.3 |
| 6 | 61–70 DPH | Gray young-2/fully fledged stage/ smoky gray | The nestlings are almost adult size and have become fully fledged. Face and beak considerably lighter in color and the overall affect is smoky gray | Though still dependant on food brought by parents, some are also seen foraging along with adults in the vicinity of the nests. Local flights have become common. An important development is that the voice is much reduced and clattering of mandibles is evident | 5.4 |
| 7 | 71 DPH to 9 months | Gray young – advanced stage | Head and neck coloration still smoky gray but retrices and remiges have turned black. Face and beak turn light yellow in color. Beak slightly decurved at the tip though not as much as in adults | Flights around the colony common. Parents still bring food to the nest, though nestlings also forage locally. After 85 DPH the gray young have become independent in all respects and feeding by parents has ceased. The voice is much reduced and clattering of mandibles common | 5.5 |

*DPH* days posthatching

**Fig. 5.2** Different stages of development of Painted Stork nestlings. (**a**) Just hatched (upto 24 h). Note that the eyes are open and the bill tip has a light (*yellow*) spot. (**b**) Between 1 and 15 DPH. The nestling is depicted sitting on its tarsi which is a very typical pose at this stage. (**c**) Between 16 and 30 DPH. Nestlings within this range are referred to as "cotton ball or snow ball." They are not yet able to stand up on their feet and still sit on their tarsi. The feather stubs become visible by this stage. (**d**) Between 31 and 45 DPH. Sometimes referred to as "turkey," at this stage the nestling is about half the adult size and can now stand on its feet. Patches of *black* begin to appear though the overall affect is still *white* and *black*. Figures not drawn to scale

*Age*: 16–30 DPH (3–4 weeks, upto 1 month)
*Weight*: 1,073–1,249 g
*Morphology*: Around 2 weeks after hatching, the nestling begins to look like what is known as "cotton ball" or "snow ball" stage. It has soft, dense wooly white down. By now the primary feather stubs are clearly discernible. Bill is black and the eye color is also black or deep amethyst. Overall, the affect is that of a considerably bigger nestling, than at the time of hatching, which is fluffy white with a jet black head, bill, and some black showing signs appearing on the body due to the development of some black retrices and remiges.
*Behavior*: The nestling is considerably noisier at this stage resorting to loud cries when hungry or at the approach of parents bringing food. Up–down display as a gesture of greeting toward parents bringing food is seen. By this stage nestlings

**Fig. 5.3** A nestling within the age range 46–60 DPH. Known as "gray young" stage, the nestling is almost three fourth the size of an adult and sometimes hops around from branch to branch and even attempts short flights. They are often seen sitting on their tarsi when not engaged in some activity. Figure not drawn to scale

are sometimes left alone in the nest by the parents. Their muscle tone has improved considerably and the body movements are perfectly coordinated. The nestlings are frequently seen flapping and exercising their wings and are able to stand on tarsi for longer periods of time, but not yet on their feet. Threat attitude is also sometimes observed among nestlings.

4. "Turkey" stage (Fig. 5.2d)
   *Age*: 31–45 DPH (5–6 weeks, approximately 1.5 months)
   *Weight*: 1,599–1,960 g
   *Morphology*: This stage is sometimes referred to as the "Turkey stage," probably because of the large size of the nestling and its movements. In terms of weight the nestling is about half the size of an adult Painted Stork. Although the general plumage is still white, but on the back it is slowly changing to black. Big patches of black are visible on the dorsal part but toward 41–45 DPH a change is visible in that the head and neck feathers begin to turn smoky gray in color. The tip of the black bill which was yellow tipped from day 1 is still visible except that the yellow patch looks slightly larger now.
   *Behavior*: As before the nestlings are still very noisy, probably noisier. Two important developments are seen at this stage. First, for the first time the nestling is able to stand up and walk on its feet in the nest. Second, their movements are quite unrestricted and the nestlings are seen exercising their wings and seen attempting to bounce off in the air. Threat displays toward nestlings in the neighboring nests are also sometimes seen.
5. Gray young-1 (Fig. 5.3)
   *Age*: 46–60 DPH (7–8 weeks, upto 2 months)
   *Weight*: 2,230–2,650 g (3/4 of adult size)
   *Morphology*: Head and neck plumage turns smoky gray and the overall affect is that of gray young. Nestlings are three fourths of adult size. The beak which was black up till now gradually starts becoming lighter in color. Remiges and retrices are well developed and the eye color is changing from amethyst to brown.

**Fig. 5.4** A nestling within the age range 61–70 DPH. The nestling looks like an adult except that it is smoky gray all over and the beak is considerably lighter in color. Figure not drawn to scale

*Behavior*: At this stage motor activity is considerably heightened and includes flapping of wings, hopping around, etc. Sometimes small flights outside the nests to nearby branches are undertaken but bouncing off in the nest is commonly observed. When the nestlings are nearly 60 days old, they start flying around in the colony. Sometimes tree-to-tree flights are also undertaken but rarely. The birds are still noisy but as days pass the bass quality of their voice gets pronounced, compared to the shrill calls produced by younger nestlings.

6. Gray young-2/fully fledged stage/smoky gray (Fig. 5.4)
   *Age*: 61–70 DPH (9–10 weeks, 2.5 months)
   *Weight*: 2,833–3,008 g
   *Morphology*: The nestlings are almost adult size and have become fully fledged. Face and beak considerably lighter in color and the overall affect is smoky gray.

**Fig. 5.5** An advanced stage of gray young, somewhere between 71 DPH to about 9 months. The face and beak are turning *yellow* (not visible in this diagram) and *black* has started to appear on the wings. Figure not drawn to scale

*Behavior*: Though still dependant on food brought by parents, some are also seen foraging along with adults in the vicinity of the nests. Local flights have become common. An important development is that the voice is much reduced and clattering of mandibles is evident.

7. Gray young – advanced stage (Fig. 5.5)
   *Age*: 71 DPH (11 weeks, approximately 3 months) to 9 months.
   This is a fairly broad stage in which the now fully grown nestling starts looking more and more like an adult.
   *Morphology*: Head and neck coloration still smoky gray but retrices and remiges have turned black. Face and beak turn light yellow in color. Beak slightly decurved at the tip though not as much as in adults.
   *Behavior*: Flights around the colony are common. Still parents bring food to the nest though nestlings also forage locally. After 85 DPH, the gray young have become independent in all respects and feeding by parents has ceased. The voice is much reduced and clattering of mandibles common.

## 5.5 Development Beyond the First Year

In the first year, the plumage starts turning white. Color of the eye becomes deep yellow (straw colored). By 16 months the birds have become voiceless like adults (Shah and Desai 1972). In their second year, the bird has acquired an almost adult like plumage but the characteristic black band across the breast is not fully developed. Also overall the body plumage is lighter than in a mature adult. However, the light yellow color of the face and beak becomes brighter. The inner secondaries of subadults do not develop the bright rosy pink color which is characteristic of adults before the second year.

## 5.6 Moult and Changes in Head Feathering

Moult in the stork family is poorly known, although it is held that there is one more or less complete moult per annual cycle occurring either after breeding (as in the case of the Marabou) or during breeding (as in the case of European White Stork) in which primaries are replaced in descendant, serially descendant, or irregular order with several active centers (del Hoyo et al. 1992). Depending upon the species, juveniles may moult into typical adult plumage during their second, third, or fourth calendar year.

One of the interesting age and breeding-related changes in the morphology of Painted Stork is the striking changes in the head plumage. The head of subadults and young adults (before reproduction) is covered with feathers (Fig. 5.6). A breeding adult however has a bald head, which has a shiny waxy skin due to lipoid secretions from the integument (Shah et al. 1977a, b).

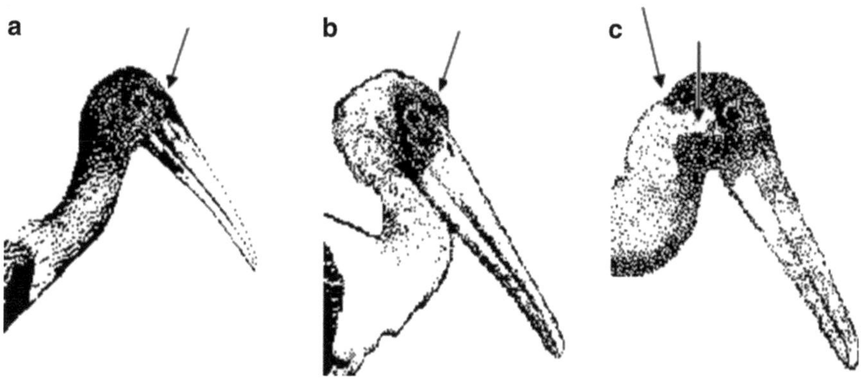


**Fig. 5.6** Changes in head feathering patterns of the Painted Stork at different stages. (**a**) Subadult, (**b**) nonbreeding adult, and (**c**) breeding adult. *Arrows* indicate the extent to which feathers occur on the head. Note that in the breeding adult, the head is completely bald and the feathers on the top of the head are pushed far back. A finger of *white* feathers is present near the ear opening, as marked by *arrow*. The difference between (**a**) and (**b**) is that while in the former the feathers are all smoky *brown* and *black*, in the adult they are *white*. Figures not drawn to scale

## References

Ali S, Ripley SD (1987) Handbook of the birds of India and Pakistan together with those of Bangladesh, Nepal, Bhutan and Sri Lanka. Oxford University Press, Delhi

Baker ECS (1932) Nidification of the birds of the Indian Empire. I. Taylor and Francis, London

del Hoyo J, Elliott A, Sargatal J (eds) (1992) Handbook of the birds of the world, vol 1. Lynx Edicions, Barcelona

Desai JH, Menon GK, Shah RV (1977) Studies on the reproductive pattern of the Painted Stork, *Ibis leucocephalus*, Pennant. Pavo 15:1–32

Kahl MP (1962) Bioenergetics of growth in nestling wood stork. Condor 64:169–183

Kahl MP (1966) A contribution to the ecology and reproductive biology of the Marabou Stork (*Leptoptilos crumeniferus*) in East Africa. J Zool (Lond) 148:289–311

Lack D (1968) Ecological adaptations for breeding in birds. Chapman and Hall, London

Meganathan T, Urfi AJ (2009) Inter-colony variations in nesting ecology of Painted Stork (*Mycteria leucocephala*) in the Delhi Zoo (North India). Waterbirds 32:352–356

Nice MM (1962) Development of behavior in precocial birds. Trans Linn Soc (N Y) 8:1–211

Rhenius CE (1907) Pelicans breeding in India. J Bombay Nat Hist Soc 17:806–807

Shah RV, Desai JH (1972) Possible explanation for voiceless ness in the adult stork, *Ibis leucocephalus*. Pavo 10:21–29

Shah RV, Desai JH (1975a) Growth and development of the Painted Stork, *Ibis leucocephala*, Pennant. I. Embryonic development. Pavo 13:83–87

Shah RV, Desai JH (1975b) Growth and development of the Painted Stork, *Ibis leucocephala*, Pennant. II. Post hatching growth pattern and motor development. Pavo 13:88–101

Shah RV, Menon GK, Desai JH, Jani MB (1977a) Feather loss from capital tracts of Painted Storks related to growth and maturity. Part 1. Histophysiological changes and lipoid secretion in the integument. J Anim Morphol Physiol 24:98–107

Shah RV, Menon GK, Jani MB, Desai JH (1977b) Feather loss from capital tract related to growth and maturity of Painted Storks. II. Distribution of glycosaminoglycans and phosphatases in the integument. Pavo 15:155–163

Starck JM, Ricklefs RE (eds) (1998) Avian growth and development. Evolution within the altricial-precocial spectrum. Oxford University Press, New York

Urfi AJ (2010) Using heronry birds to monitor urbanization impacts: a case study of Painted Stork *Mycteria leucocephala* nesting in the Delhi Zoo, India. Ambio 39:190–193

# Chapter 6
# Food and Foraging

*It [Painted Stork] stalks about the shallows with its bill in the water, partially held open, and instantly seizes any fish, frog, or crab that comes in its way. If the fish be a spiny one, it crushes its spines between its strong mandibles, and swallows it, head foremost. During the heat of the day, it stands motionless in water, knee deep, digesting its morning meal.*

T.C. Jerdon (1864)

**Abstract** The Painted Stork mostly feeds by tactolocation and its bill is adapted for this purpose. The possible significance of a slight downward curvature of the bill, foraging behavior, nocturnal foraging, and kleptoparasitism is discussed in this chapter. The foraging habitats of Painted Stork are mostly shallow marshes and wetlands. It also frequents coastal sites but it remains to be established how much of visual foraging component is involved. A comparative study of foraging behavior with the Milky Stork is included which, it is believed, segregates ecologically and is more partial to coastal sites than the Painted Stork. This chapter also includes the impacts of Painted Stork and other heronry birds on their ecosystem, in terms of fish harvesting and enrichment of the waters due to their droppings.

## 6.1 Introduction

Foraging ecology is one of the most advanced areas of modern Ornithology. While studies on small waders, chiefly belonging to families Scolopacidae, Charadriidae, etc. are numerous, those on large waders are few and far between. However, the Wood Stork has been quite well studied and some classic studies include an elucidation of its tactile foraging mode (Kahl and Peacock 1963; Kahl 1964), prey size selection (Ogden et al. 1976), foraging habitat characteristics (Coulter and Bryan 1993), seasonal variations in its diet, foraging behavior and flock sizes (Gonzalez 1997), etc. Among Asian storks the foraging behavior of the Black-necked Stork, which is also a fish-eating bird like the Painted Stork, has been studied in relation to

A.J. Urfi, *The Painted Stork: Ecology and Conservation*,
DOI 10.1007/978-1-4419-8468-5_6, © Springer Science+Business Media, LLC 2011

seasonal water level fluctuations in India (Maheswaran and Rahmani 2002). Regarding the Painted Stork, besides information on its diet (Ali and Ripley 1987; Desai 1971; Desai et al. 1974) and foraging habitats (Sundar 2006), a detailed study on its foraging behavior in the Delhi region was undertaken by Kalam and Urfi (2008). Some information about the diet and foraging behavior of the Milky Stork is also available (Li et al. 2006; Swennen and Marteijn 1987), which is discussed from a comparative viewpoint in this chapter.

## 6.2 Foraging Behavior

Jerdon's observation, reproduced at the beginning of this chapter, captures the essence of Painted Stork's foraging behavior, though the part about the bird crushing spiny fishes before swallowing has probably not been observed very often. However, many species of wading birds are known to mandibulate their prey with bites or bill snaps (Kushlan 1978). The bills of storks are fairly hardy structures, and so it may not be entirely improbable that they crush fishes having spines before swallowing them.

Field observations indicate that Painted Storks forage largely by tactolocation, though the visual mode is also sometimes employed (Fig. 6.1). Before we move on to discuss the forging behavior of Painted Stork, it is important to understand what is tactile foraging, and how do we know that a stork is feeding by this mode? The pattern of tactile foraging is broadly the same in all *Mycteria*, in that the foraging individual inserts its partially open bill in the water, sometimes moving it from side to side, and walks forward. Foot stirring is used and sometimes the wing is flashed open, presumably to startle the concealed fish into the open mandibles. The eyes always remain above the water mark. Since the water in which the bird is foraging is often muddy or hazy, most probably the bird cannot see the prey very clearly.

Tactile foraging works best if the density of prey is high, because then the encounter rate with prey is also high. In the case of the Wood Stork, which is essen-

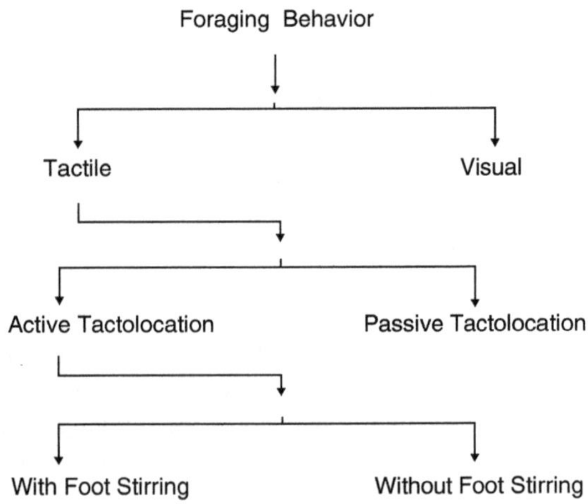


**Fig. 6.1** Schematic representation of foraging behaviors exhibited by Painted Stork

tially a dry season nester, the high prey density effect is produced when waters dry up leading to a concentration of fish in shallow pools. According to some estimates, densities of fish may be as much as 8,000 fish/m$^2$ in drying pools of Florida (del Hoyo et al. 1992). In the case of Painted Stork, the high prey densities result from entirely different factors, viz. the glut of prey soon after the monsoon rains due to the spawning of fish. Painted Stork seem to time their nesting with this seasonal abundance of prey. An idea about how high the prey densities are can be had from some data gathered during the course of the hydrobiological project of BNHS at Keoladeo Ghana National Park, Bharatpur (Ali and Vijayan 1983). At one of the sluice gates, regulating the entry of water inside the park, nearly 336 fish fry were obtained in a plankton net of 1 m$^2$, operated for 85 s. In other words, four fry per second were entering the park through an area of 1 m$^2$. Thus an estimated 8,600,000 fish fry would be entering the park in 24 h.

Kalam and Urfi (2008) noted that of the 323 fish captured by 103 different individuals, 96% were caught by the tactile mode. However, in one case, at Bhindawas Bird Sanctuary near Delhi, storks were observed to be opportunistically picking up dead fish floating on the surface of water. So, while tactile foraging seems to be the norm, visual foraging is also used, albeit sparingly and opportunistically.

### *6.2.1 The Trophic Apparatus*

The organ of primary interest from the viewpoint of forging is of cource the bill, although it may have other uses too, such as pulling out twigs during nest construction and as a weapon during territorial fights with rivals, at the nesting and foraging grounds (see Chap. 3). A few things are immediately apparent if we examine a Painted Stork bill at close hand. It is large, more than 24 cm, waxy yellow, and has a slight curvature at the tip (Fig. 6.2). The Painted Stork does not hold its bill completely horizontal to the ground, either in flight or while standing thus giving the impression that it has a strongly decurved bill. Actually, the bill is straight for most of its length but at a point roughly 185 mm from the mouth, a slight decurvature of approximately 11° is recorded (A.J.Urfi unpublished observations). A slight gap, measuring less than 1 mm, is also observed between the mandibles in a locked

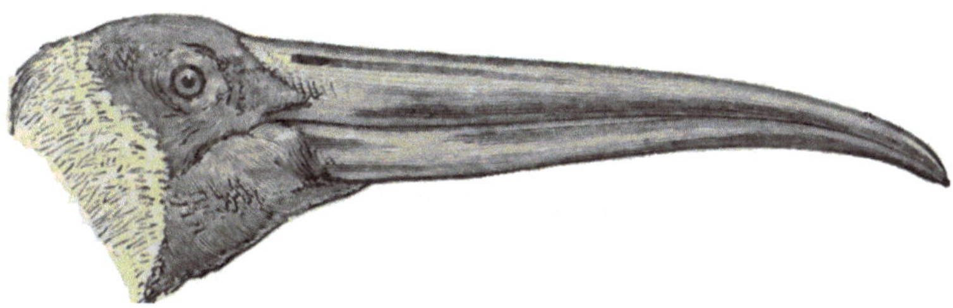

**Fig. 6.2** The head and bill of the Painted Stork. *Source*: Blanford (1898)

position. The line along which the upper and lower mandibles are locked is sinuous, with an elevation in the distal portion and a depression in the proximal part. Thus, the intermandibular gap is variable along the length of the bill, and the inner surfaces of the mandibles are more or less horizontally placed in relation to each other at some points. The functional significance of this feature is discussed below.

The questions that arise in context of the functional morphology of Painted Stork bill are basically two. First, how does the bill perform tactolocation and second why is it slightly decurved? Regarding the first, in an elegant experimental setup to study the reflex action of the mandibles in the Wood Stork, Kahl and Peacock (1963) demonstrated that the lower mandible snaps shut as soon as it meets any object. The time elapsed between the first contact and capture was estimated to be only 25 ms, making it one of the fastest reflexes recorded in the vertebrate world (del Hoyo et al. 1992). The bill has some tactoreceptors in the mandibles although it is not clear exactly where they are located since no histological study has been performed. However, it seems that the bill snap reflex works independently of any receptors.

The Kahl and Peacock model successfully explains many aspects of tactile foraging in *Mycteria* but there are also some problems with it (Kushlan 1978). For instance, the mandibles snap shut when the jaw muscle is activated by a sudden jolt, say a fish hitting it. However, while forging in a vegetated area, it would be impossible to avoid the mandibles meeting some object accidentally, say a plant. Yet we do not see the bill needlessly snapping shut every now and then. Also when moving through tangled vegetation, the bird does not seem to get confused between fish and plants. del Hoyo et al. (1992) suggest that fish over 3.5 cm are preferably taken, so clearly this is a system which is partly selective. How can this happen in a system which is based on tactoreception?

To test if the Wood Stork could feed in total darkness by tactolocation, Kahl and Peacock took their studies a step further. For this purpose, they blindfolded their experimental storks by fixing ping-pong balls over their eyes to cut off vision temporarily. They discovered that blindfolding had no effect and the storks could forage as well as in normal condition. This explains why the system is so effective in muddy waters and theoretically speaking why the Painted Stork should have no difficulty foraging in darkness.

Now how do we reconcile with the slightly decurved bill of the Painted Stork? Although down curving bills are not very common in the bird world, being restricted to ibises, kiwis, flamingos, curlews, sunbirds, and some species of hummingbirds, their functional significance has prompted several investigations. Restricting ourselves only to large waders, in the case of Common Curlew, it was first hypothesized that the bill decurvature had arisen by selection favoring either a greater search-arc for detection of prey at the wintering grounds or an easier capture of insects among vegetation on the breeding grounds (Owens 1984).This view was challenged by Davidson et al. (1986) who argued that the decurvature was most probably an adaptation for a particular technique for prey capture. Comparisons with a straight-billed wader Bar-tailed Godwit, which feeds along with the curlew on mudflats, suggested that the decurved bill helped in the intact removal of long prey such as polychaete worms. A straight bill was not successful in removal of prey intact with the result

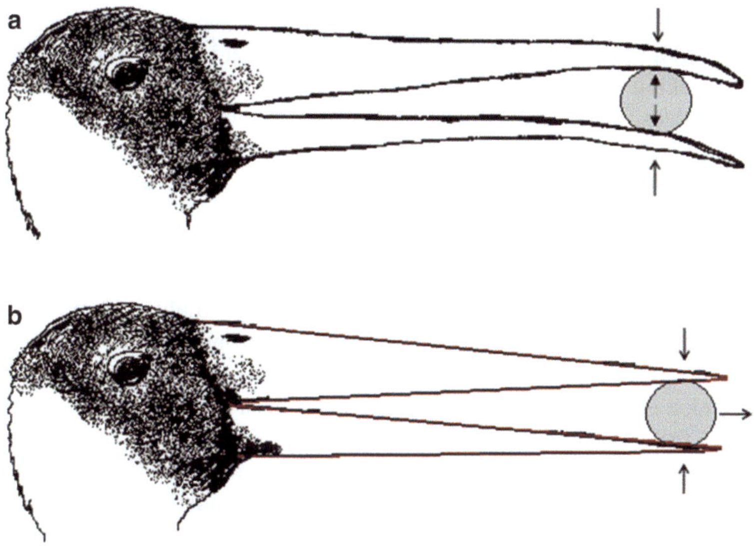


**Fig. 6.3** (**a**) A Painted Stork holding a prey item at the tip of its beak. As the mandibles will press the fish, being soft will get flattened and a larger surface of it will come in contact with the bill. Due to the decurved shape of the bill, when the mandibles open at the tip their surfaces will be more or less parallel to each other, thereby permitting greater contact with the prey. (**b**) Depicts the situation in a hypothetical straight bill. As the mandibles press against each other, the resulting pressure will tend to push the prey forward and it may fall off. Also a much lesser surface area of the mandibles comes in contact with the prey and so the grip will be weak

that much of the nutrients in the body of the prey remained in the mud and thereby unattained by the predator.

Another large wader with a decurved bill, which has been the subject of a detailed investigation, is the ibis. In the coastal states of Southern USA, the White Ibis forages on Mud Fiddler Crabs (*Uca pugnax*). The ibis probes in the mud for crabs, which live in burrows connected to the surface by a tunnel (Bildstein 1993). Interestingly, this tunnel is curved in a manner which almost mimics the shape of the bill of ibis. Since ibises are observed to rotate around openings of crab burrows with their bills inserted in the tunnels, it was initially suspected that this action could help in aligning the bill along the curvature of the burrows in order to reach the crabs easily. However, no evidence was found in favor of this hypothesis. By comparing the size of the gap along the length of the bill, it became apparent that in the case of a decurved bill there would be a more horizontal spread of the mandibles in contrast to a straight bill, in which the gap between mandibles would progressively decrease from the distal to the proximal end. It was argued that in the former case a stronger grip on the prey would result simply because more area would come in contact with the prey (Bildstein 1993). Also, if the bill had a different shape, say straight, then there would be a tendency for the prey to be pushed toward the tip from where it may easily fall off (Fig. 6.3).

As far as the Painted Stork is concerned, they too are sometimes observed to pivot around a point, much as the ibises do. They do this more often in vegetated habitats, making extensive use of foot stirring to flush the prey toward their partly open mandibles (Kalam and Urfi 2008). Whether they do this to orient the position of the bill for any particular purpose is not clear. In any case, their prey is mostly fish, which seldom buries itself in the mud (except for some catfishes and perhaps some other fishes) in the manner of invertebrates like worms and crustaceans that constitute the diet of other waders.

One of the reasons why the bill of the Painted Stork is decurved at the tip, which is also probably the area that is most sensitive, could be simply to obtain a purchase on the prey. On the other hand, the curvature at the tip and the sinuous interlocking along the length of the bill could also be an adaptation for obtaining a strong grip on a prey, which is liable to easily escape after it has been captured. Anybody who has held a live fish in his hand would know that it slips out easily due to powerful jerks of its body. A layer of mucous on the surface of the fish also makes it very slippery. Its powerful jerks are produced by contraction of nonoxidative, fast, white muscle fibers of the myotomal mass (Bone 1978). In our observations it was rare that fish, once it had been caught, escaped from the mandibles, and in those cases in which large fishes were dropped, the action seemed deliberate. Indeed the few fishes dropped were generally those which were very large (24–32 cm) and expectedly had a long handling time (1–6 min) (Kalam and Urfi 2008). Thus, selection that enhances the grip on the prey is likely to be a strong driving force behind the evolution of bill shape and possibly the slightly down curved bill of the Painted Stork could be an adaptation for this purpose.

How do bill shapes vary among other storks and to what extent are they a reflection of the diet or foraging method? The decurved bill is a characteristic of the genus *Mycteria* and all the four species under this genus are primarily piscivorous. They forage in muddy waters where catching fish by using visual cues works in a limited way. Indeed, the Milky Stork which consumes large mudskippers is recorded to drag its bill in the wet mud in an attempt to make contact with its prey, in a manner strongly reminiscent of other *Mycteria*. But, all *Mycteria* also opportunistically take other types of prey items, as and when encountered, probably by employing visual cues. For instance, the Wood Stork consumes crayfish, amphibians, insects, small snakes, and even baby alligators (del Hoyo et al. 1992). The Painted Stork has been recorded trying to swallow a snake (Urfi 1988), which could have been located by sight.

Other than *Anastomous*, which do not enter this discussion directly since they have a specialized diet, most members of tribes *Ciconiini* and *Leptoptilini* have straight bills. *Ciconia* have a characteristic visual foraging method and a catholic diet comprising a number of vertebrate and invertebrate organisms. Some species like the White Stork are long-distance migrants and exhibit considerable variations in their diets at their nesting and wintering grounds. *Ephippiorhynchus* have straight, thick, and slightly upcurved bills. These storks are primarily fish eaters and catch prey by walking in the shallows and stabbing the bill repeatedly in water, clearly suggesting that they employ a visual approach. However, *Epiplatys senegalensis* is observed foraging in muddy waters, catching prey by tactolocation in a manner similar to *Mycteria* (del Hoyo et al. 1992) and *E. asiaticus* is also known to employ a tactile foraging mode (Maheswaran and Rahmani 2002). But in these storks, large or awkward fish is generally brought to dry ground where it is stabbed and swallowed. By contrast, *Mycteria* do all their feeding while in the water itself.

Jabiru has a straight but slightly upturned bill. While it is primarily a fish eater, its foraging mode is varied comprising tactile (35%), visual (37%), and mixed (28%) components. All species of *Leptoptilos* have straight, thick, and slightly upturned bills and forage visually, on fish and a variety of other vertebrate and invertebrate items, and also on offal at garbage dumps.

Examining different images of Painted Stork, one is struck by the variations in the size and shape of the bill. In some cases, the bill looks decidedly thin and long, like a sword while in others it appears thicker and shorter like an elongated wedge. It is well known that the tropic organ in birds is a highly plastic structure. In some birds, such as the European Oystercatcher, which exhibits different types of foraging modes and prey choices, the variations in bill shape are extremely high and fixed in different individuals (Hulscher 1996). In the case of Painted Stork, it is unlikely that the variations in its bill morphology would be so enormous. Although many morphological variations could simply be tied down to age and sex differences, more studies in this direction would be extremely useful.

## 6.3 Diet of the Painted Stork

Two separate studies, both from the Delhi region, North India, have tended to confirm that the Painted Stork is a predominantly fish-eating bird. First, Desai et al. (1974) examined the gut contents of adult and nestling storks (five individuals of each) at the Delhi Zoo colony and reported the presence of local fish, mostly of commercial significance. These were *Wallago attu* ($n=4$), *Puntius amphibius* (3), *Catla catla* (3), *Cirrhina mrigala* (2), *Mystus seenghala* (2), *Channa punctata* (2), and one each of *Heteropnestes fossilis*, *Channa striata*, *Labeo rohita*, *Notopterus chitala*, and *Notopterus notopterus*. Besides the above, fragments of plant fibers and other plant material including leaves, small stones, aquatic insects, and a small frog were also recorded. Since all species of storks are almost exclusively carnivorous, the plant matter found in their guts is likely to be that taken by accident when the prey is caught (del Hoyo et al. 1992) and therefore unlikely to be of much significance.

In Kalam and Urfi's (2008) study since neither gut content analysis nor any field examination of prey items was undertaken, it was not possible to ascertain the species of fish captured. In just 13 cases where the identification of prey species was possible, the results were *W. attu* ($n=3$), *C. punctata* (6), *Mastacembalus armatus* (1), and *Tilapia mossambica* (3). Interestingly, the last two species were not recorded in the earlier study from Delhi (Desai et al. 1974). *Tilapia* is an exotic fish which was introduced in India in 1952 (Jhingran 1982) and has since then become a major invasive species throughout the region. Its presence in the diet of Painted Stork in 2004 and absence during the study undertaken in late 1960s could be indicative of fluctuations in its relative abundance over the decades.

Besides the ubiquitous fin-fish, other organisms in the diet of Painted Stork have also been reported, such as reptiles, frogs, crustaceans, and insects (Ali and Ripley 1987). At Kokkare Bellur colony in South India, adults were observed feeding the nestlings with frogs, crabs, and large insects like grasshoppers (Sridhar et al. 2002). While some other storks, Black-necked Stork and Adjutant, have been recorded catching birds (Panday 1974), nothing so spectacular has ever been reported about the Painted Stork, though on one occasion, as already mentioned, an individual was observed trying to swallow a live water snake.

Large numbers of Painted Stork along with flamingo, egret, reef heron, and other waders are observed foraging in the salt pans around Meethapur in Jamnagar district of Gujarat (A.J. Urfi, unpublished observations) as well as at other coastal sites in India (Parasharya and Naik 1990). It is difficult to accept that they would not be feeding on crustaceans and other invertebrates, which are common in such waters and therefore in such situations visual foraging should be employed more often than tactolocation.

## 6.4 Field Studies in the Delhi Region

Using a videographic approach the foraging ecology of Painted Stork was studied in detail at 14 different sites scattered within a radius of about 180 km around Delhi. The study included a diverse range of habitats, including marshes, both riverine and seepage pond type. The size of prey was quantified by methods standardized beforehand as detailed in Kalam and Urfi (2008).

Painted Storks do not feed continuously throughout the day. Typically, they forage in bouts followed by long periods of rest. The length of the bouts was recorded to range from 5 to 60 min (mean $23.65 \pm 17.64$, $n = 17$). While no significant difference between bout duration in the forenoon and afternoon periods was recorded, interestingly, of the 127 bouts recorded in our study, 6% were those which were abruptly terminated by disturbance due to a human or cattle. Interestingly, all of the disturbed bouts were recorded outside the protected areas. The conservation implications are discussed later (Chap. 8).

In our study Painted Stork mostly foraged in a manner which can be termed as "active tactolocation," i.e., the foraging bird walked in the water with its head immersed. However, in less than 10% of all video recordings, the foraging individual also stood stationary and waited for the fish to make contact with the mandibles. This, so-called "passive tactolocation" (Fig. 6.1) would be expected to occur more at sites where the water was flowing but actually it was equally likely to happen at both lentic and lotic sites.

Foot stirring which is commonly observed in *Mycteria* has also been observed in Snowy Egret (Willard 1977; Master et al. 1993) and Little Egret (Erwin et al. 1985). It is believed to serve the function of dislodging prey that is hidden in the vegetation

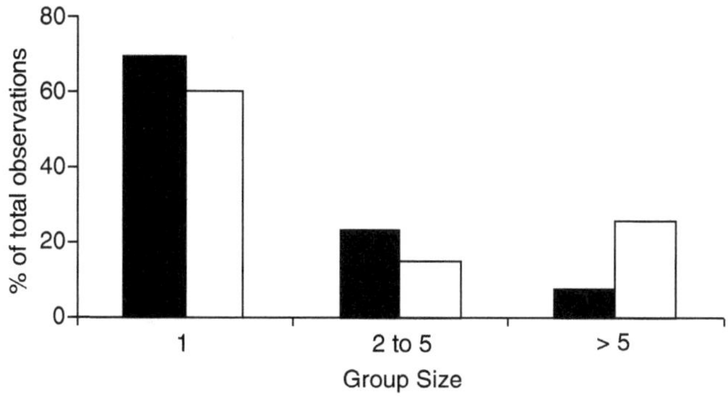


**Fig. 6.4** Foraging group size of Painted Stork in breeding (*filled squares*) and nonbreeding (*open squares*) seasons in Delhi in 2004–2006. *Source*: Kalam and Urfi (2008)

or buried in the mud. Foot stirring was observed in 75% cases. The foot-stirring rates, expressed as number of times the foot-stirring behavior was observed in 5 min intervals, in the vegetated habitats (mean $51.2 \pm 26.6$) were significantly higher (Mann–Whitney *U*-test, $P < 0.001$) compared with nonvegetated habitats (mean $27.25 \pm 27.09$). Interestingly, foot-stirring rate was negatively related to group size in both types of habitats. The likely explanation for this could be that due to the presence of several birds at one spot, the fish may be disturbed anyway (Master et al. 1993), thereby making additional foot stirring redundant for flushing out hidden prey. That the foot-stirring rate in vegetated habitats was higher than the nonvegetated ones suggests that in the former there could be a greater need for it.

One line of reasoning would suggest that since Painted Storks are colonially nesting birds, they would also be flock foragers (Brown and Brown 2001). This is because there is a strong suggestion that colonies also serve as information centers about location of food resources. At the Delhi Zoo, Painted Storks are observed to fly to their foraging grounds, chiefly Okhla Bird Sanctuary in small groups but once having reached the foraging patch they disperse. In our study, Painted Storks were observed foraging singly or in groups which ranged from 2 to 18 individuals (mean $2.78 \pm 3.34$, $n = 185$). The group size in the breeding season (mean $2.22 \pm 3.05$, $n = 125$) was significantly smaller than the group size in the nonbreeding season (mean $3.93 \pm 3.63$, $n = 60$) (Mann–Whitney *U*-test, $P < 0.001$; Fig. 6.4). This could be due to several factors. For instance in the nonbreeding season, i.e., summer, as the ponds and wetlands dry up, birds tend to concentrate on a few remaining ones.

### 6.4.1 Prey Sizes Captured and Regurgitated

A comprehensive body of theory, Optimal Foraging Theory (OFT), has been developed to explain and predict foraging behavior in animals (Krebs and Davies 1984; Stephens and Krebs 1986). While one set of OFT models attempts to predict the

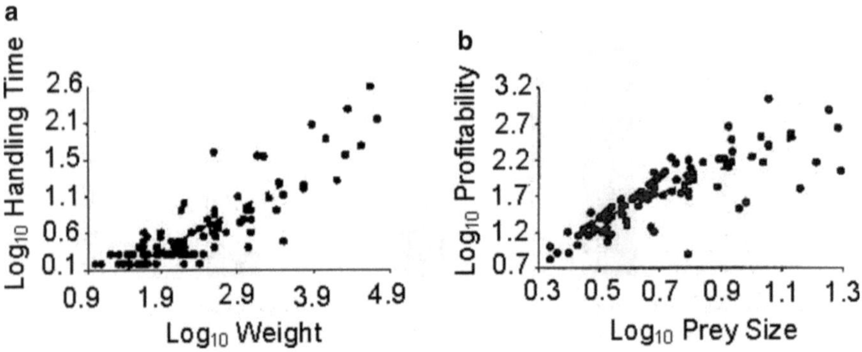


**Fig. 6.5** (**a**) Relationship between prey weight (mg) and handling time (s) and (**b**) prey length (cm) and profitability (mg/s). Note the scale is $\log_{10}$ in both graphs. *Source*: Kalam and Urfi (2008)

time a foraging animal should spend in a food patch ("patch models"), another set attempts to predict the prey sizes which a forager would be expected to capture, purely on the basis of energetic considerations, if given a choice ("diet models").

On examining the relationship between prey size and handling time (*ht*) using a selection of video images, we found, expectedly, a relationship between the two. As Fig. 6.5a shows, with increasing prey size the *ht* increases. Small prey are just swallowed and negligible time is spent in handling it or orienting it properly. However, small prey does not bring much energetic returns because of its low energy content (*E*). In other words, its profitability is low since

$$\text{Profitability} = \frac{\text{energy content } (E)}{\text{handling time } (ht)},$$

where $E$ is in energy units (kJ) and $ht$ is time (s or min).

While it would be expected that the forager would try to catch bigger prey, beyond a point it would not be profitable to do so (Fig. 6.5b). In our study very large fish, >20 cm, were occasionally caught. However, there are also cases on record when Painted Stork after having caught a large fish dropped it (Sankhala 1990). We observed that on several occasions Painted Stork after having caught a large fish gave up midway or after prolonged lengths of time (high *ht*).

Most of the fish captured ranged between 1 and 12 cm. In the breeding season, the dominant size class was 1–6 cm, in complete contrast to those captured in the nonbreeding season when the dominant size class was >12 cm (Fig. 6.6). It is difficult to say if these differences are due to active selection by the Painted Stork or simply a reflection of the abundance of these size groups at this time of the year. For instance in the months September onward, the dominant size class for most species of fishes would be small since they have emerged from the breeding activity of fishes during the monsoon (Jhingran 1982). In the summer, however, large-sized fishes, mostly adults, which survived the heavy predation pressure earlier on, by all

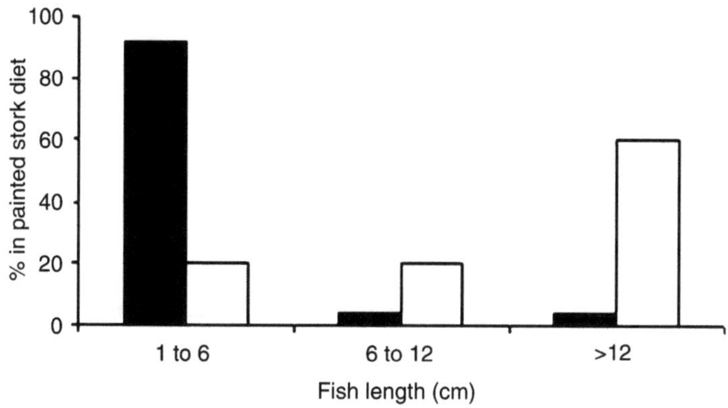


**Fig. 6.6** Length of fishes taken by Painted Stork at different foraging sites in breeding (*filled squares*) and nonbreeding (*open squares*) seasons in Delhi in 2004–2006. *Source*: Kalam and Urfi (2008)

manner of fish-eating birds including the Painted Stork, would out class very small-sized fishes.

The usual method of feeding nestlings in all *Mycteria* as well as other colonial waterbirds is by regurgitation on the nest floor. As the parents alight at the nest from a food-finding mission, the hungry nestlings immediately start clamoring for food in their shrill voices. After much begging and calling, the parent bird finally regurgitates the contents of its crop onto the nest floor. In our study we found most of the prey items regurgitated by the parent birds were fish, although in some cases, when the nestlings were very small (<15 days) the parent birds were observed regurgitating a gooey mass (bolus), which could possibly be semi-digested fish. It is held that the crop in the case of Painted Stork has some digestive functions and is known to predigest the food (Ali and Ripley 1987).

The number of fish regurgitated in a single regurgitation bout ranged between 1 and 13 (mean $2.91 \pm 2.27$). About 80% of the fish regurgitated at the nest were smaller than 10 cm (Fig. 6.7). In some cases it was observed that if the parent birds regurgitated fish >20 cm, these were not eaten by the nestlings. Interestingly, the prey length (mean $5.92 \pm 2.49$ cm, $n = 167$) regurgitated at the time when the nestlings were very small (<30 days old) was significantly lower from the prey length (mean $6.99 \pm 3.77$ cm, $n = 178$) regurgitated a few weeks later, when the nestlings had become older (>30 days) (two-tailed paired *t*-tests, $P < 0.01$; Fig. 6.7).

## 6.5 Comparisons with Milky Stork

While in many respects there are similarities in the foraging behavior of Painted Stork and its congener Milky Stork, there are also several differences. Much like the Painted Stork, the Milky Stork also employs a tactile foraging technique, involving

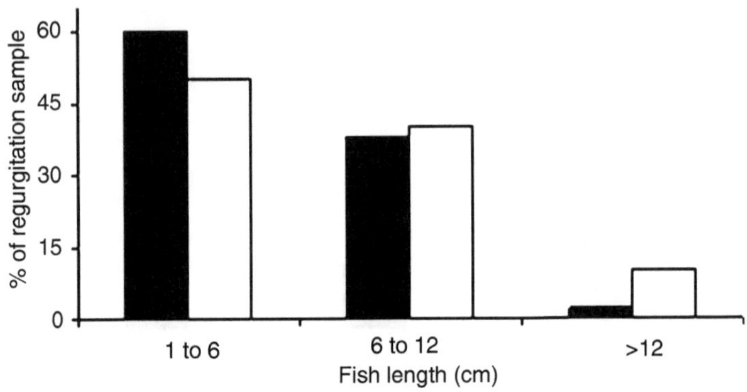


**Fig. 6.7** Length of fish regurgitated to younger (<30 days) and older (>30 days) nestlings by Painted Stork at the Delhi Zoo during 2004–2006. *Black bars* (*filled squares*) show younger and *white bars* (*open squares*) represent older nestlings. *Source*: Kalam and Urfi (2008)

standing still or walking through mud or shallow water. It probes in the mud with a partly opened bill, often drawing it in an arc from side to side and employing foot stirring, until a prey item is located (Swennen and Marteijn 1987; Silvus and Verheugt 1989; Indrawan et al. 1993). It has been recorded feeding in loose flocks with a gap of about 50 to 100 m between individuals (Li et al. 2006). Sometimes Milky Storks also move in a single tight flock, flushing fish in shallow water (Indrawan et al. 1993) like the Painted Storks do. Milky Storks often feed in aggregations with other wading birds, such as Lesser Adjutant and egrets (Verheugt 1987; Hancock et al. 1992) as does the Painted Stork, although it is more likely that in both cases such aggregations are determined by chance and there is no real association among different species as such.

Among the points of difference between the two, the Milky Stork is perhaps more likely to indulge in nocturnal foraging in comparison to the Painted Stork. For instance, at Pulau Dua, Indonesia, considerable nocturnal activity was noted, with birds both foraging and visiting nests during hours of darkness, at least under a full moon (Hancock et al. 1992). It has been suggested that the two species separate ecologically, with the Painted Stork frequenting fresh water habitats and the Milky Stork being partial to coastal sites. Significantly, a recent study from the Sumatra province of Indonesia reported the diet of Milky Stork as consisting of Milkfish (*Chanos chanos*), Elongate Mudskipper (*Pseudapocryptes elongatus*), Giant Mudskipper (*Periaptholomodon schlosserii*), and Mullet *Moolgarda* sp./*Chelon* sp., many of which are coastal or brackish water species (Iqbal et al. 2009). Also, at coastal sites food availability in the intertidal zone is strongly governed by the tidal cycle, rather than the diurnal cycle and this makes for a strong case for nocturnal foraging in the Milky Stork.

While searching for food on mudflats, the Milky Stork is known to look for mudskipper holes and probe its bill in and around the hole up to 10–15 times, sometimes immersing the whole bill and head in the mud. When the mudskipper is caught, it is hauled out and oriented properly, head first, and then swallowed. Another, rarer alternative involves the bill being inserted into a hole and pushed forward, opening up a groove in the mud. Any prey which comes in contact with the bill is immediately captured as this method is entirely tactile (del Hoyo et al. 1992). Such highly specialized behaviors have not been observed in the Painted stork.

## 6.6 Variations in Foraging Activity

As mentioned previously, Painted Storks typically forage in bouts of varying durations. Bouts of feeding are followed by periods of inactivity when the bird either stands on one leg or hunched up or is seen sitting on bent tarsi on the shore. The period of inactivity could be due to the fact that though the bird is satiated and its gut is full, internally its digestive system is working hard, digesting the food and expending energy in the process (the so-called "hidden foraging time"). Alternatively, the periods of rest could be dictated by factors linked to prey availability and accessibility.

On account of several factors, there is generally a huge gap between what is available for a foraging bird and what is eventually taken by it and so generalizations often tend to oversimplify the picture. For instance, not all species of fish are equal in terms of their energy or nutrition content. Different species of fishes have different calorific values, based on their chemical composition (Love 1980). Also, different species of fishes exhibit different behaviors which influence their harvestability by predators. Water depth might be one criterion. A classical example, though not entirely relevant in the case of wading birds like Painted Stork, is the vertical stratification among the three major carps of India *L. rohita*, *C. mrigala*, and *C. catla* with one inhabiting the surface, another the water column, and the third being a bottom dweller. More importantly, there are significant differences in the activity patterns of different fishes (Jhingran 1982). To take a broad example, while most carps are diurnal, many catfishes particularly the Indian Catfish (*H. fossilis*) are nocturnally active.

Painted Storks are believed to forage actively in the early hours of the day and become inactive during the afternoons, resuming foraging again in the late afternoon or evening (del Hoyo et al. 1992). Since at different times of the day, the oxygen content of the water changes due to fluctuations in surface temperature, this is likely to influence prey behavior. Thus during early morning, when the dissolved oxygen content of waters is low, fish may come to shallow areas or close to the surface or make more frequent trips to the surface for gulping air and in the process be easily caught. At other times of the day, they may go into the deeper areas, out of the reach of the storks. In this respect, more studies on the foraging activity patterns of Painted Stork are warranted.

Age and sex-specific differences in the foraging abilities of different individuals in a population are also known. In our study also, some differences in the foraging

**Table 6.1** Mean number (±SD) of attempts, capture and foraging success of Painted Stork during 5 min observation periods, with respect to season and age, in the Delhi region in 2004–2006

| Age | *n* | Attempts/5 min | Prey items captured/5 min | Foraging success |
|---|---|---|---|---|
| Adult (breeding) | 123 | 80.5 ± 28.1 | 1.66 ± 2.84 | 0.03 ± 0.05 |
| Adult (nonbreeding) | 24 | 57.1 ± 32.2 | 0.62 ± 0.49 | 0.03 ± 0.04 |
| Juvenile (nonbreeding) | 36 | 53.47 ± 17.6 | 0.50 ± 0.65 | 0.01 ± 0.02 |

*Source*: Kalam and Urfi (2008)

abilities of different types of birds were observed (Table 6.1). Juvenile birds in the nonbreeding season had a lower foraging success compared to adults. This could be due to lack of experience in very young birds.

## 6.7 Nocturnal Foraging

While the Milky Stork seems to indulge in nocturnal foraging quite often, little is known about the ability of Painted Stork to forage at night. Unfortunately, this aspect has been largely ignored in the case of Painted Stork. Workers who have taken it upon themselves to investigate nighttime foraging in waders lament that this is a general case since most investigators, themselves being daytime workers, are partial toward diurnal species (Sitters 2000). However, one case of night foraging by the Painted Stork on record is in the form of some general observations recorded by Kannan and Manakadan (2007) at Pulicat Lake in South India. In 2005 due to the activity of some local fishermen, Painted Storks were disturbed and resorted to resting in nearby abandoned crop fields adjacent to the lake in the company of other birds. After 19:00 h, when human disturbance was less, the birds were observed to come back for foraging in the lake. In fact, on several days in July 2005 Painted Storks were observed feeding at the lake between 19:00 and 23:00 h. While these observations suggest that Painted Storks are capable of nocturnal forging, it also seems that they resort to it only when they have been unable to meet their required intake targets during the day and are very hungry. Since the authors of the above-cited report mention that the birds had been disturbed preceding their night foraging schedule, it would seem that these birds were "time stressed." Foraging for longer to make up for the lost foraging time is something which has also been reported from studies on other waders (Urfi et al. 1996).

## 6.8 Kleptoparasitism

Food stealing is quite common in birds. Why take the trouble of catching fish when it can be easily stolen from a conspecific or some other bird. In our study (Kalam and Urfi 2008) many birds were observed feeding alongside Painted Stork.

If we discount those species with which any type of interaction with the Painted Stork is impossible, due to differences in prey choice (such as Spoonbill, ibis, Openbill, etc.) and focus only on fish-eating waders of a comparable body size, then there are only two species Grey Heron and Great Egret in which a Painted Stork can possibly take some interest from the point of view of stealing fish. The former was recorded to co-occur with the Painted Stork on 32% of all observations while the latter on 25% of all observations.

Only 22 aggressive interactions were recorded during our study of which 14 were intraspecific and 8 were interspecific. Painted Storks have been recorded to also steal fish from other piscivorous birds, i.e., nonwaders. Mahindiran and Urfi (2010) recorded an instance of Painted Stork stealing fish from a Little Cormorant. Although the fish taken by the Little Cormorant is small (57.45 ± 3.42 mm), since this species forages near the shore in shallow areas, it was possible for the Painted Stork to steal fish from it. This is in contrast to Indian Cormorant and Large Cormorant, which feed on larger fish but are partial toward deeper areas of the wetland where the Painted Stork cannot reach. Black-necked Stork, being of a larger size, have been recorded to attack Painted Stork and steal fish from them (Naoroji 1989).

While at the foraging site the Painted Stork is able to bully most birds and steal prey, except the Black-necked Stork, at the nesting site it is often at the receiving end. According to some reports, at Keoladeo Ghana, Bharatpur, egrets on hearing the clamoring calls of young ones begging for food alight in the vicinity of Painted Stork nests or those chicks resting on mounds, waiting patiently, ready to snatch a morsel with a lightning dash should an opportunity arise (Naoroji 1989). During a study in 1985–1986, the egrets were sometimes observed to fly in from neighboring blocks on hearing Painted Stork young begging for food. For instance, once they were observed to fly from block B to L, which is a sizeable distance. Naoroji (1989) notes that in spite of so much effort the egrets were rarely successful in grabbing a beakful. The egrets were almost totally dependant upon their own hunting ability, yet persisted with this activity and as soon as the Painted Stork had finished feeding their young ones, they promptly flew away from the area. Interestingly, other observers who also made observations on egrets stealing food at Painted Stork nests in Keoladeo had a different take on this. For instance, Bhatnagar and Raizada (1979) observed that egrets used to "pick fish from different sides [of the nest] as crows pick feed from a tray," implying that they were able to steal food with ease.

## 6.9 Foraging Habitats

It is interesting to visualize the threat perceptions of fish from birds, at a generalized Indian wetland. Such an attempt has been made in Fig. 6.8 from which it is clear that fish are very persecuted creatures indeed and face attacks from all sides. Birds of prey such as Osprey, Pallas Fish Eagle, and fish-eating owls can scoop up any fish that come to the surface for breathing. The surface is also under attack from birds such as skimmer, terns, gulls, kingfisher. Pelicans can harvest fish in large shoals at

**Fig. 6.8** Representation of how different types of fish-eating birds distribute themselves by virtue of their adaptations and feeding specializations in a generalized wetland. The vertical scale shows depth in feet. The birds shown are *1* Osprey; *2* skimmer; *3* kingfisher; *4* darter; *5* pelican; *6* Painted Stork; *7* heron; *8* cormorant

**Table 6.2** Foraging depth of Painted Stork recorded in the Delhi region

| Foraging depth (cm) | *n* | % of observation |
|---|---|---|
| 1–12 | 113 | 34.98 |
| 12–25 | 196 | 60.68 |
| 25–43 | 14 | 4.33 |

*Source*: Kalam and Urfi (2008)

just below the surface. Most of the deep parts of the wetland are accessible to diving birds such as cormorants and darter. However, in all this the Painted Stork is only one threat and is restricted only to the shallow parts of the wetland.

Unlike a diving bird, say a cormorant which has access to fish resources in the pelagic zone of the wetland, the Painted Stork is restricted to the littoral zone. Instances of swimming have not been recorded in the Painted Stork or in other storks. The Painted Stork can feed in water only as deep as its feet will take it. In the Delhi region we observed that in about 60% cases, Painted Stork foraged in water 12–25 cm deep (Table 6.2). On fewer occasions (35%) they were recorded foraging in shallower areas (1–12 cm deep) and on only about 4% occasions in deeper waters (>25 cm).

By and large, all species of birds segregate themselves by size, if they belong to similar niches. In the littoral zone too, body size and length of the legs would be the deciding factor. Comparatively speaking, while smaller species like herons can access a very shallow zone, Painted Stork can go into slightly deeper waters.

By and large, Painted Stork is mostly an inland species and seeks freshwater marshes and shallow wetlands. Both lentic and lotic sites are used (Kalam and Urfi

2008). For its type of adaptations, deep reservoirs and lakes, except for their shores, are out of reach and therefore meaningless. In India, Painted Storks are also sometimes seen foraging in agricultural fields, probably looking for insects, other invertebrates, frogs, and lizards. However, where opportunity exists, Painted Storks utilize coastal areas for foraging. However, studies by Parasharya and Naik (1990) showed that while Painted Stork nesting in gardens of Bhavnagar city frequented coastal wetlands and salt pans (25% each of all observations), there was decidedly a preference for freshwater marshes (50%). One of the significant factors in context of taking marine food is the excessive amount of salt that may get into the body of the predator along with the food. However, fish-eating birds are probably not be restricted by salt stress because their principal prey – fish – themselves osmoregulate in marine habitats. The little salt that does get ingested along with the food may be a problem but most adult colonial waterbirds are capable of excreting excess salt through an infraorbital salt gland, although young birds in some species are not very efficient in getting rid of the salt that comes with food, especially marine crustaceans (Frederick and Spalding 1994).

## 6.10 Painted Stork in an Ecosystem Context

Living organisms do not exist in isolation; as much as they are affected by their environment, they in turn influence their environment in direct and indirect ways and in varying degrees. About the role of birds in general, Wiens (1973) commented,

> .....it seems unlikely that birds exert any major influence on ecosystem structure, functional properties, or dynamics through their direct effects on either the flux or storage of energy or nutrients....their role instead might be as governors or controllers...

However, many instances of impact of aquatic birds on their ecosystem are well known. One important role which birds play is in transporting seeds, spores, etc. of algae and other aquatic organisms, externally, i.e., as having them on feet and feathers and bill, and internally, i.e., in feces (Kristiansen 1996). It has been suggested that the plankton diversity of most lakes is similar because of dispersal by migratory birds on a global scale.

Birds which exclusively feed on vegetation, whether submerged, free floating, or growing on the shore of wetlands, such as waterfowl, remove a substantial portion of the standing crop. Grazing on aquatic macrophytes by migratory geese and their impact on crop removal has been studied at Keoladeo Ghana, Bharatpur (Middleton and Vander Valk 1987) and the role of birds in influencing the dynamics of wetlands highlighted. Planktivorous birds, such as flamingos, remove plankton and in the process impact the plankton community structure directly (Hurlbert and Chang 1983). Fish-eating birds, by removing fish, also remove the predators of plankton and therefore they influence the community structure indirectly.

One of the significant roles of colonially nesting waterbirds is to enrich the waters by their droppings. One of the earliest studies in this regard estimated that the role

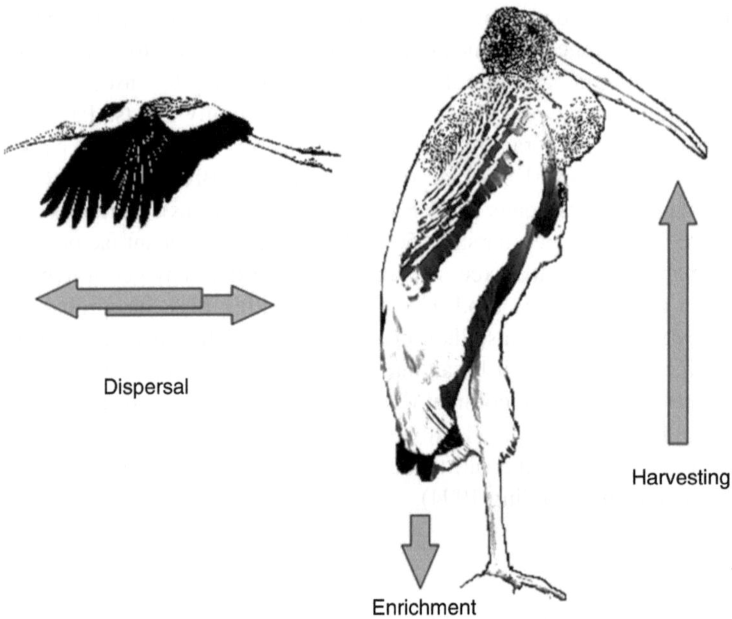


**Fig. 6.9** The possible ways in which Painted Stork interacts with its environment

of aquatic birds, chiefly cormorant, pelican, and other sea birds in cycling phosphorous in a global context, was quite significant (Hutchinson 1950).

In context of their wetland ecosystem, Painted Storks constitute a tiny element of fish harvesters. So overall, any ecological impact that they can have on their ecosystem – whether it be freshwater marshes or the coastal areas – should be quite limited. However, the possible ways in which Painted Stork and other species of large waders can influence their habitat (Fig. 6.9) are harvesting and enrichment as discussed below.

### *6.10.1 Harvesting and Enrichment*

The foraging activity of Painted Stork results in the removal of standing crop. But do we have any idea about how much quantity of fish, of what species or size groups, are removed by predation? Taking consideration of BMR into account, Desai et al. (1974) estimated that a colony of about 100 nests of Painted Stork would consume about 24 metric tons of fish during one breeding season (i.e., a 150-day period). This estimate was derived largely from studies undertaken at the Delhi Zoo and the basic unit was a family consisting of two adults and 1.1 chicks. For Keoladeo Ghana, Ali (1953) using different criteria (2,000 nests, family size equal to two adults and two chicks but a 90-day period) came up with an estimate of 90 tons. According to him since 2,000 nests were representative of only a small fraction of the colonies at Bharatpur, the fish harvest figures for the entire park would be much greater.

Both of the above estimates pertain to only one species, i.e., Painted Stork, and if other species of piscivorous birds are included then the total harvesting would be much greater. Also, the estimates above pertain only to the nesting season, i.e., 90–150 days. If a whole year is taken into account, then the amount of biomass (fish) removed by predation would be many times higher than the above estimates. Naturally, most of the fish taken during the breeding season would be small-sized fish (1–8 cm, Fig. 6.6) though in the nonbreeding season Painted Stork would be expected to take larger fish. What consequences this will have on fish population structure and indirectly on aquatic communities need to be carefully examined. Given the sheer quantity of biomass that is consumed, the chances of toxins in the food chains particularly pesticides and their biomagnification are enormous. Therefore, Painted Storks, being placed at the highest trophic level in their ecosystem, are likely to be affected to a very large extent.

While foraging, Painted Storks are likely to enrich the water by their droppings. However, since the foraging individuals are spaced out, the effects of their droppings are likely to be close to being insignificant. However, at the nesting colonies, the droppings of birds make an impact due to the sheer numbers of nesting individuals involved. Also, in the mixed species heronries, there would be not just the Painted Stork but several other species as well. In this regard, the only study on record seems to be the one conducted by Paulraj (1988) at Vedanthangal Bird Sanctuary in Tamil Nadu. A chemical analysis of the reservoir water, rich in the droppings of colonial waterbirds, indicated that it contained 0.373 ± 0.01 mg/L ($n=9$) of nitrite ($NO_2$), 0.199 ± 0.012 mg/L ($n=9$) of nitrate ($NO_3$), and 17.4 ± 1.3 mg/L ($n=9$) of phosphorus ($P_2O_5$).

### *6.10.2 Other Possible Effects*

Many large waders such as ibis and Wood Stork tend to disturb the substrate while foraging. This process, known as "bioturbation," tends to mix up the constituents (Bildstein 1993). While Painted Stork and other waders are foraging in the water, it is likely that spores of microorganisms, plankton, and seeds of plants will get stuck on their body feathers, feet, or even bill and be transported to other sites. Although no study has been done as such it may be useful to explore what is the role of birds like the Painted Stork in this context.

## References

Ali S (1953) The Keoladeo Ghana of Bharatpur (Rajasthan). J Bombay Nat Hist Soc 51:531–536

Ali S, Ripley SD (1987) Compact handbook of the birds of India and Pakistan. Oxford University Press, Delhi

Ali S, Vijayan VS (1983) Hydrobiological (ecological) research at Keoladeo National Park, Bharatpur (first interim report). Bombay Natural History Society, Mumbai

Bhatnagar RK, Raizada M (1979) Feeding competitor of nestlings of Painted Stork. Newslett Birdwatchers 8:12–13

Bildstein KL (1993) White ibis. Wetland wanderer. Smithsonian Institution Press, Washington

Blanford WT (1898) The fauna of British India, including Ceylon and Burma. Birds, vol IV. Taylor and Francis, London

Bone Q (1978) Locomotor muscle. In: Hoar WS, Randall DJ (eds) Fish physiology, vol VII. Academic, New York

Brown CR, Brown MB (2001) Avian coloniality, progress and problems. In: Nolan V Jr et al (eds) Current ornithology. Kluwer Publishers, New York

Coulter MC, Bryan AL Jr (1993) Foraging ecology of wood storks (Mycteria americana) in east-central Georgia. I. Characteristics of foraging sites. Colon Waterbirds 16:59–70

Davidson NC, Townshend DJ, Pienkowski MW, Speakman JR (1986) Why do curlews *Numenius* have decurved bills? Bird Study 33:61–69

del Hoyo J, Elliott A, Sargatal J (eds) (1992) Handbook of the birds of the world, vol 1. Lynx Edicions, Barcelona

Desai JH (1971) Feeding ecology and nesting of painted stork Ibis leucocephalus at the Delhi zoo. Int Zoo Yearbook 11:208–215

Desai JH, Shah RV, Menon GK (1974) Diet and food requirement of painted stork at the breeding colony in the Delhi zoological park. Pavo 12:13–23

Erwin RM, Hafner H, Dugan P (1985) Differences in the feeding behaviour of little egrets (*Egretta garzetta*) in two habitats in the Camargue, France. Wilson Bull 97:534–538

Frederick PC, Spalding MG (1994) Factors affecting reproductive success of wading birds (Ciconiiformes) in the Everglades ecosystem. In: Davis S, Ogden J (eds) Everglades: the ecosystem and its restoration. St. Lucie Press, Delray Beach

Gonzalez JA (1997) Seasonal variation in the foraging ecology of the wood stork in the Southern Llanos of Venezuela. Condor 99:671–680

Hancock JA, Kushlan JA, Kahl MP (1992) Storks, ibises and spoonbills of the world. Academic, London

Hulscher JB (1996) Food and feeding behaviour. In: Goss-Custard JD (ed) The Oystercatcher from individuals to populations. Oxford University Press, Oxford

Hurlbert SH, Chang CCY (1983) Ornitholimnology: effects of grazing by the Andean flamingo (*Phoenicoparrus andinus*). Proc Natl Acad Sci U S A 80:4766–4769

Hutchinson GE (1950) The biogeochemistry of vertebrate excretion. Bull Am Museum Nat Hist 96:554

Indrawan M, Lawler W, Widodo W, Sutandi (1993) Notes on the feeding behaviour of Milky Storks *Mycteria cinerea* at the coast of Indramayu, west Java. Forktail 8:143–144

Iqbal M, Takari F, Mulyono H, Rasam (2009) A note on the breeding success of Milky Stork *Mycteria cinerea* in 2008, South Sumatra province, Indonesia and more on its diet. BirdingASIA 11:73–74

Jerdon TC (1864) The birds of India, vol III. George Wyman & Co., Calcutta

Jhingran VG (1982) Fish and fisheries of India. Hindustan Publishing Corporation, Delhi

Kahl MP (1964) Food ecology of the wood stork (*Mycteria americana*) in Florida. Ecol Monogr 34:97–117

Kahl MP, Peacock LJ (1963) The bill-snap reflex: a feeding mechanism in the American Wood Stork. Nature 199:505–506

Kalam A, Urfi AJ (2008) Foraging behaviour and prey size of the painted stork (*Mycteria leucocephala*). J Zool (Lond) 274:198–204

Kannan V, Manakadan R (2007) Nocturnal foraging by Painted Stork *Mycteria leucocephala* at Pulicat Lake, India. Indian Birds 3:25–26

Krebs JR, Davies NB (eds) (1984) Behavioural ecology: an evolutionary approach. Blackwell Scientific Publications, London

Kristiansen J (1996) Dispersal of freshwater algae – a review. Hydrobiologia 336:151–157

Kushlan JA (1978) In: Sprunt A, Ogden JC, Winckler S (eds) Feeding ecology of wading birds. National Audubon Society Research Report No. 7. National Audubon Society, New York

Li ZWD, Yatim SH, Howes J, Ilias R (2006) Status overview and recommendations for the conservation of Milky Stork *Mycteria cinerea* in Malaysia: final report of the 2004/2006 Milky Stork field surveys in the Matang Mangrove Forest, Perak. Wetlands International and the Department of Wildlife and National Parks, Peninsular Malaysia, Kuala Lumpur, Malaysia

Love RM (1980) The chemical biology of fishes. Vol. 2: advances 1968–1977. Academic, London

Maheswaran G, Rahmani AR (2002) Foraging behaviour and feeding success of the black-necked stork (*Ephippiorhynchus asiaticus*) in Dudwa national park, Uttar Pradesh. India J Zool (Lond) 258:189–195

Mahinderan M, Urfi AJ (2010) Foraging patterns and kleptoparasitism among three sympatric cormorants (*Phalacrocorax* spp.) from the Delhi region, North India. Hydrobiologia 638:21–28

Master TL, Frankel M, Russell M (1993) Benefits of foraging in mixed-species wader aggregations in a southern New Jersey salt marsh. Colon Waterbirds 16:149–157

Middleton B, Vander Valk AG (1987) The food habits of Greylag and Barheaded geese in the Keoladeo National Park India. Wildfowl 38:94–102

Naoroji R (1989) Unusual interactions for food. J Bombay Nat Hist Soc 86:100

Ogden JC, Kushlan JA, Tilmant JT (1976) Prey selectivity by the wood stork. Condor 78:324–330

Owens NW (1984) Why do Curlews have curved beaks? Bird Study 31:230–231

Panday JD (1974) Storks preying on live birds. J Bombay Nat Hist Soc 71:141

Parasharya BM, Naik RM (1990) Ciconiiform birds breeding in Bhavnagar city, Gujarat. In: Daniel JC, Serrao JC (eds) Conservation in developing countries: problems and prospects. Bombay Natural History Society, Mumbai

Paulraj S (1988) Impact of guano deposition in Vedanthangal Waterbird Sanctuary (Chengalpattu district, Tamil Nadu). J Bombay Nat Hist Soc 85:319–324

Sankhala K (1990) Gardens of God. The Waterbird sanctuary at Bharatpur. Vikas Publishing House, New Delhi

Silvus MJ, Verheugt WJM (1989) The status of storks, ibis and spoonbills in Indonesia. Kukila 4:119–132

Sitters HP (2000) The role of night-feeding in shorebirds in an estuarine environment with specific reference to mussel-feeding oystercatchers. DPhil Thesis submitted (Hillary Term 2000), University of Oxford, Oxford

Sridhar S, Chakravarthy AK, Srihari K, Prakash NA (2002) Priorities for conservation of Kokre Bellur heronry in Karnataka, South India. In: Rahmani AR, Ugra G (eds) Birds of wetlands and grasslands: proceedings of the Sálim Ali centenary seminar on conservation of avifauna of wetlands and grasslands. Bombay Natural History Society, Mumbai

Stephens DW, Krebs JR (1986) Foraging theory. Princeton University Press, Princeton

Sundar KSG (2006) Flock size, density and habitat selection of four large waterbirds species in an agricultural landscape in Uttar Pradesh, India: implications for management. Waterbirds 29:365–374

Swennen C, Marteijn ECL (1987) Notes on the feeding behaviour of the milky stork (*Mycteria cinerea*). Forktail 3:63–66

Urfi AJ (1988) Painted Stork *Mycteria leucocephala* (Pennant) swallowing a snake. J Bombay Nat Hist Soc 86:96

Urfi AJ, Goss-Custard JD, Durell SEA (1996) The ability of oystercatchers *Haematopus ostralegus* to compensate for lost feeding time: field studies on individually marked birds. J Appl Ecol 33:873–883

Verheugt WJM (1987) Conservation status and action programme for the Milky Stork (*Mycteria cinerea*). Colon Waterbirds 10:211–220

Wiens JA (1973) Patterns and process in grassland bird communities. Ecol Monogr 43:237–270

Willard DE (1977) The feeding ecology and behavior of five species of herons in Southeastern New Jersey. Condor 79:462–470

# Chapter 7
# Storks and Humans

**Abstract** This chapter discusses the cultural context of storks as well as people's perceptions about heronries and storks in general. The White Stork occupies a prominent place in Western culture, often as a symbol of fertility and good omen. In Sanskrit literature of ancient India, there is a mention about several traits of storks such as their dance and courtship rituals, voicelessness, and their massive bills. In contemporary India, storks are often depicted in paintings and postage stamps. In certain tribal communities, like Mirshikars of North India, ceremonies revolving around the Black-necked Stork are known to exist.

## 7.1 Storks in World Traditions

Large-sized birds, regularly seen in a certain localities and possessing interesting behavioral traits, tend to be easily noticed and therefore figure prominently in folklore and cultural traditions. Storks have long, strong bills, and this lends itself easily to symbolism of the phallus and, associated with it, fertility. Combined with the quiet, industrious nesting activity exhibited by a breeding pair and the focus of their activities on rearing chicks and raising a family, storks also symbolize domestic concerns and values. Interestingly, storks themselves are not generally regarded as "sexy," in the sense of having panache. Other birds, perhaps birds of prey or the pheasants with their bold, brilliant plumages seem to possess this trait, but not storks.

The White Stork features in several myths, stories, tales, and beliefs in Western or European traditions. For instance, in ancient Greece it was regarded as sacred to the Goddess Hera, the protector of nursing mothers. This could probably be the origin of the Western folk belief that storks bring babies. The stork's association with the fertile waters of creation also echoes the story of it bringing new born babies into the world. According to an ancient legend, children are embryos in the cosmic waters,

A.J. Urfi, *The Painted Stork: Ecology and Conservation*,
DOI 10.1007/978-1-4419-8468-5_7, 

**Fig. 7.1** A cardboard cutout model of a stork with a bundle, bearing a baby (a plastic doll), held in its beak. This model was displayed outside a home, in a locality in Saarbruicken, South Germany, where a baby had been recently born. Photo: A.J. Urfi (1997)

and are discovered by the stork as it searches for fish (Saunders 1995). All in all, the bird is widely regarded as a symbol of fertility across Europe and often depicted on greeting cards with a bundle, bearing a new born baby, held in its beak. Whether this is, as Kahl (1971) puts it, a "…cute and convenient way of avoiding early sex education," even today in many countries across continental Europe, particularly Poland and Germany, cardboard or wooden models of storks with a bundle in its beak are kept outside houses where a child has been recently born (Fig. 7.1).

The stork figures prominently in Aesop's fables. In Christian lore, it is regarded as the heralder of spring and therefore linked to purity, piety, and resurrection. It is popular not just in its breeding ground, i.e., Europe, but also across considerable parts of its wintering range, particularly in North Africa and Asia. Across many different cultures, the White Stork has a remarkably consistent set of positive and benevolent symbolic attributes. In China, for instance, the stork is associated with filial piety because it is believed to care for its aged progenitors as well as own offspring. Another connection, widespread in the East, is with longevity which in Taoism becomes extended to immortality (Saunders 1995).

## 7.2 Storks in the Indian Context

Ornithological and natural history observations preserved in ancient Indian texts such as the Vedas, Puranas, etc., written in Sanskrit and Pali, provide fascinating insights not only into the world of nature but also into the manner in which ancient Indians perceived nature. Some of it can be glimpsed in the "Panchtantra" and "Jataka" tales, which are stories of wit and wisdom, conferring human attributes to animals and birds, widely acknowledged to be the original source of Aesop's Fables (Urfi 2008).

One of the most significant contemporary compilations of literature on birds from ancient India, dating back to ca. 1000 BC and before, is the book entitled "Birds in Sanskrit Literature" by Dave (1985). From this book, where a full chapter has been devoted to storks and their allies, one learns that the Woolly-necked Stork and Black-necked Stork figure in the famous Indian epic "Mahabharata." This is when Arjuna, the hero of the epic, speaks of the great speed of his chariot drawn by white steeds and makes a reference to the soaring storks. From Dave's work we also know that several characteristics of storks, such as the details of their dancing and courtships displays, were recorded which are available today in the form of verses and poems. One of the characteristics of storks observed by the ancients was their voicelessness. This finds expression in an ancient Sanskrit verse, highlighting the virtues of silence,

> The parrot and the mynah speak according to the good or bad qualities that they possess the stork, on the other hand, does not speak and (its) silence can be interpreted any which way.

Dave's work unearths several interesting references to the scavenging habits of the adjutant storks, their massive bills looking like "plough shares." The Woolly-necked Stork and its habit of "standing motionless on one leg as if absorbed in meditation" (Ali and Ripley 1987), also find a mention in the texts of ancient India.

### *7.2.1 Storks in Tribal Customs: The Mirshikars*

In India, storks figure in the traditions of some tribal communities, particularly those who have been traditionally associated with hunting and trapping of birds and animals. One such community is "Mirshikars," who live in the North Indian state of Bihar and adjoining areas. These people are Muslims by religion and have traditionally been hunters and trappers for several generations. Being experts at handling birds and wildlife, they also have a highly developed vocabulary. The Painted Stork figures in their vernacular by the name of *Kăkāri* (Ali and Ripley 1987). An interesting report by Grubh and Sekhar (1968) details the marriage and courtship customs of Mirshikars. According to this report, if a Mirshikar youth wanted to marry a girl, besides fulfilling all other conditions and formalities, he would have to catch an adult Black-necked Stork (locally known as "Loha Sarang"), alive, single handedly and unarmed, but for his glue and rod device. Thankfully, this bizarre custom ended long ago due to a causality which occurred when a Black-necked Stork severely injured a prospective bridegroom.

### 7.2.2 *Storks in Folk Art*

Pictorial representation of birds and animals is a common element of Indian folk art, both traditional and modern. One of the best examples seen every day, especially while traveling on interstate highways, is the highly stylized drawings of birds on goods ferrying trucks. Although depictions of storks are difficult to come by, an image of an egret is commonly seen on most trucks, suggesting that Ardeids, close relatives of storks, do make an impact on people's minds (Fig. 7.2). However, occasionally, one finds storks represented in paintings by local artists. An image of the Painted Stork in the form of a painting on a wood panel originating from Karnataka state is shown in Fig. 7.3. Given that Karnataka is an important place for Painted Stork and the famous Kokkare Bellur heronry is located in this state, it is not surprising that local artists include it in their work.

**Fig. 7.2** Stylized drawing of an egret on a goods ferrying truck

**Fig. 7.3** A painting on a wood panel depicting Painted Stork roosting on trees. This decoration piece was photographed at an exhibition of Karnataka state handicrafts. Photograph: A.J. Urfi (2001)

### *7.2.3 Storks on Postage Stamps*

The depiction of any object, person, or place on a postal stamp or currency notes is reflective of their importance and significance. In this regard, storks seem to score rather well as they appear on stamps and currency notes of several countries. All the species of *Mycteria* have been represented on postage stamps but restricting ourselves only to the Asian species, the Milky Stork figures on the postage stamps of Indonesia and Malaysia. The Painted Stork appears on a currency note of Sri Lanka and on postage stamps from Vietnam, Thailand, and Cambodia.

The Indian Postal department has brought out a number of stamps depicting waterbirds. There are three stamps which bear the image of Painted Stork. Three stamps on just one species of stork appear to be an unusually high number. It would almost seem as if the defining characteristic of the Indian postal department is an "inordinate fondness" for the Painted Stork. The first among the Painted Stork stamps (Fig. 7.4) was released in 1976 to mark the celebration of Keoladeo Ghana Bird Sanctuary. Set against the background of a blue sky, the stamp shows one Painted Stork flying off, probably on a food-finding mission. The second stamp, a se-tenant,[1] was issued in 1996 to mark the birth centenary of Dr. Salim Ali, the legendary Indian ornithologist and conservationist, who was also instrumental in persuading the Maharaja of Bharatpur to convert his private hunting ground Keoladeo Ghana into a national protected area. Salim Ali was very closely connected with this site and initiated a number

[1] A se-tenant, composite of two stamps of different denominations, is a special stamp issued on rare and special occasions.

**Fig. 7.4** Facsimile of the stamp showing Painted Stork issued by the Indian Postal department in 1976 to mark the occasion of creation of Keoladeo Ghana Bird Sanctuary. The Hindi name of the bird "Janghil" is mentioned in Devanagari (Hindi) script at the *top*

**Fig. 7.5** Facsimile of the se-tenant postage stamp issued by the Indian Postal Department in 1996 to mark the birth centenary of Dr. Salim Ali. The scene depicted in the stamp is that of Keoladeo Ghana National Park in Bharatpur, which has extensive colonies of the Painted Stork (as shown in the picture)

**Fig. 7.6** Facsimile of the stamp on the Painted Stork issued in 2000. In this stamp, as also in the 1976 stamp (Fig. 7.4) the Hindi name for the Painted Stork "Janghil" is mentioned at the *bottom*. It is not clear why the artist has chosen to depict the bird perched on a branch and not in the water

of ornithological activities here, particularly the bird ringing and migration project and the hydrobiological project on behalf of the BNHS. While one part of the se-tenant stamp has a portrait of Salim Ali, as he looked in his later years, the other part shows a Painted Stork nesting colony at Bharatpur (Fig. 7.5). The third stamp, issued in 2000, is a drawing of Painted Stork in rust-colored hues. The Hindi name for the bird "Janghil" is clearly mentioned in the Devanagari script, on the stamp (Fig. 7.6).

## 7.3 People's Attitudes Toward Heronries

It is widely held that benign, tolerant attitudes toward wildlife, birds, and environment prevail in many parts of India. The tradition of "sacred groves," small patches of forests devoted to local deities, in some localities is well known (Gadgil and Vartak 1975). Local people's positive sentiments toward birds and animals are believed to have benefited many wildlife species, such as the Sarus Crane (Gole 1989) and Blackbuck (CEE 1997). With respect to heronries, similar positive and protective sentiments have been recorded by ornithologists and naturalists.

As already discussed, the heronries of the village Kokkare Bellur are an important breeding ground for both Painted Stork and Spot-billed Pelican. For several decades, these birds have coexisted with man and generally there has been no molestation to the birds by local villagers. According to Manu and Jolly (2000),

> Though they [the villagers] do not attribute any divine status to the birds, they have always offered them protection, believing that the birds bring them good fortune with regard to the rains and their crops. They are proud of their long association with the birds, nicknamed 'daughters of the village', and compare them to the local girls who may marry into another village but inevitably return home to deliver and nurse their newborn babies.

The villagers' protective attitudes toward the birds nesting in the village are highlighted by several incidents mentioned in the above report. Once, when some outsiders (called "hakki-pikki" tribals) tried to steal eggs and nestlings of storks and pelicans from nests, they were arrested by the villagers and asked to pay a fine. When these intruders did not pay up, they were locked up in a room and released after some time. In another instance in the early 1970s, when a Bangladeshi refugee attempted a similar nest-robbing feat, he was caught and locked up for a day in the school.

### *7.3.1 Koonthangulam Narratives*

Koonthangulam Bird Sanctuary, the site of a village heronry in Tamil Nadu state of South India, has been visited and revisited by generations of naturalists and much has been written about people's sentiments vis-a-vis the heronries. Also known by other names such as Kanthakoolam, Kundukulam, Koondakulam, etc., the site lies in Tirunelveli district, about 20 km from Tirunelveli town. It is an important place from the view point of nesting Painted Stork, pelican, and other colonial waterbirds (Islam and Rahmani 2008). Basically, the heronry is located near a rain and river fed tank in the village precincts, with a number of other smaller tanks in the vicinity. One of the earliest references to this colony is by Rhenius (1907), who wrote

> The villagers look on these birds as semi-sacred and will not allow anyone to disturb or molest them, so they return to build there year after year, and have done so for years past.

While this captures the positive sentiments of local people toward the heronry, Webb-Peploe (1945), writing some decades later, suggests that the colonies could also have a nuisance value associated with them

> The noise and the smell caused some of the villagers to suggest destroying the nests and driving the birds away some years ago, but the head-men of the village protect them.

Noise and smell are certainly irritating aspects of bird colonies but still it is amazing that the villagers were reported to regard these birds in such high esteem (semi-sacred). A couple of decades later Wilkinson (1961) writes

> The headmen of the village still protect the birds and their women- folk spoke with scorn of a village of which they had heard where the people had so ill-treated their birds that 'not even a sparrow is to be found there now!

Interestingly, a decade later Abraham (1973) adds a twist to the story

> Unlike the villagers of Kundukulam in Nangunery of Tirunelveli District, who propose to destroy the pelican colony because of the noise and bad odor produced by the pelicans and Painted Storks, the villagers in the Kanjirankulam very zealously guard the colony, even to

> the extent of caring for the young which fall from their nest due to wind or some other cause. The villagers feed these forlorn young with a meal of fish, frogs and snakes which are chopped up for the purpose. I have been told that even the village dogs do not molest these hapless young chicks which slowly grow and finally join their tribe and flyaway.

From the above it is clear that whether it be Koonthangulam or some other site in India, most authors of natural history have written in glowing terms about the local people's inclination and interest in preserving heronries. It is another matter altogether that the interest exhibited could also be a slight exaggeration. Most villagers, being a marginalized lot economically as well as socially and so being at the receiving end of the state administrative apparatus, both in British India as well as Independent India, may perhaps be more inclined toward giving an answer that points toward the positive features of heronries. ("All is well" is probably a safer answer than to say "All is not well; the birds are a nuisance; the trees should be cut down," etc.) Also, marginalized farmers and laborers in villages harbor a degree of suspicion toward functionaries of the "state," for which the reasons are well known. Probably, in British times a police officer masquerading as a bird lover or ornithologist was as much an object of loathe as today the city-bred birdwatcher, dressed in western clothes is. Perhaps, the villager senses that the bird watcher or bourgeoisie conservationist is eager to hear only the good news, which is that the local villager/tribal is there to protect the birds that he likes so much. As long as the local is there, it does not matter if in abject poverty, the birdwatcher cum environment crusader can rest assured that his beautiful, harmonious, natural world is intact.

Having said this let us also look at some aspects of the local people's love for the heronries. That villagers regard the coming of storks and pelicans as "harbingers of good times and good crops" is understandable. Birds have an uncanny ability to sense rain and their arrival at the nesting site is a sure sign that the monsoon that year will be a good one and this, to the tiller of the soil, translates into economic boom and a promise of some degree of prosperity.

As a resource, heronries provide meat and eggs and if need be local people will steal eggs and poach birds when they have to. But one of the most significant aspects of heronries is the enormous quantities of fertilizers they generate by the droppings of birds (Paulraj 1988). In case of heronries where nests are located on islands, the droppings directly fall in the water while in the case of village heronries, where nests are built on tall trees the droppings fall on the ground. Most authors have tended to mention the guano benefits too in their reports which bring us to the question, do these apparently protective sentiments also have practical and material reasons behind them, as well.

While many other writers have waxed eloquent about the protective attitudes of locals, observations by an officer in British Indian Army Captain Bates are quite straightforward. Bates (1931) is forthright when writing about the local people's sentimental attitudes toward the heronries. He treats the subject in a practical manner and chooses to take a direct route, taking cognizance of "mercenary reasons"

> ….I fear me a love of bird –life does not enter into the picture: the reason is purely a mercenary one. Owing to the water of the tank being considered to possess high fertilizing properties, the value of the land irrigated by it is assessed for revenue purposes at a higher rate than any other land in the vicinity.

## References

Abraham S (1973) The Kanjirankulam breeding bird sanctuary in the Ramnad district of Tamil Nadu. J Bombay Nat Hist Soc 70:549–552

Ali S, Ripley SD (1987) Compact handbook of the birds of India and Pakistan. Oxford University Press, Delhi

Bates RSP (1931) Birdlife in India. Madras Diocesan Press, Madras

Centre for Environmental Education (CEE) (1997) Biodiversity (enviroscope, a manual for college teachers). World resources Institute USA and CEE. Oxford University Press, New Delhi, Ahmedabad

Dave KN (1985) Birds in Sanskrit literature. Motilal Banarsidass, Delhi

Gadgil M, Vartak VD (1975) Sacred groves of India – a plea for continued preservation. J Bombay Nat Hist Soc 72:314–320

Gole P (1989) The pair beside the lake. ICF Bugle 15(4):4–5

Grubh PR, Sekhar PB (1968) Blacknecked Stork (*Xenorhynchus asiaticus*) and the marriage of Mirshikars. Newslett Birdwatchers 8:1–2

Islam MZ, Rahmani AR (2008) Potential and existing Ramsar sites in India. Bombay Natural History Society, Mumbai

Kahl MP (1971) The courtship of storks. Nat Hist 80:36–45

Manu K, Jolly S (2000) Pelicans and people: the two-tier village of Kokkare Bellur, Karnataka, India. Community based conservation in South Asia: case study no. 4. Kalpavriksh and International Institute of Environment and development, New Delhi

Paulraj S (1988) Impact of guano deposition in Vedanthangal Waterbird Sanctuary (Chengalpattu district, Tamil Nadu). J Bombay Nat Hist Soc 85:319–324

Rhenius CE (1907) Pelicans breeding in India. J Bombay Nat Hist Soc 17:806–807

Saunders NJ (1995) Animal spirits. Macmillan, London

Urfi AJ (2008) Birds of India – a literary anthology. Oxford University Press, New Delhi

Webb-Peploe CG (1945) Notes on a few birds from the south of the Tinnevelly district. J Bombay Nat Hist Soc 45:425–426

Wilkinson ME (1961) Pelicanry at Kundakulam, Tirunelveli district. J Bombay Nat Hist Soc 58:514–515

# Chapter 8
# Conservation

*Earth provides enough to satisfy every man's need, but not every man's greed.*

Mahatma Gandhi

**Abstract** This chapter starts by examining the evidence for the suspected decline in Painted Stork populations throughout India, with a focus on some selected nesting colonies. Threat factors, ongoing conservation work, and suggestions for future work, including launching long-term conservation monitoring programs are highlighted in this chapter.

## 8.1 Is the Painted Stork Declining?

A maxim in conservation biology rules that large sized animals are more prone to endangerment as compared to smaller ones. The Painted Stork, which is a fairly large bird, does not seem to be an exception. It is feared to be in danger and is currently listed as near threatened (BirdLife International 2001). However, thankfully, the bird is not seriously endangered yet, although it is plagued by a host of problems throughout its range. Its foraging habitat i.e., wetlands are severely threatened and in some places its meat and eggs are eaten and so the bird is persecuted. The fact that the Painted Stork is a carnivorous bird does not make things any easier for it because being at the apex of aquatic food chains it is vulnerable to bioaccumulation of toxins and pesticides in its body. One of the surest signs of decline of any species is when fewer individuals are recorded year after year from sites which are its regular haunts (Urfi et al. 2005). However, since we do not have any long-term data on the populations of Painted Stork we have to resort to scattered reports and anecdotal records, some of which indeed seem to suggest that all is not well for heronry birds in general and Painted Stork in particular.

Site-specific studies, sustained over a long period of time provide the best picture about the status of any species. In India such studies are few and from the information

A.J. Urfi, *The Painted Stork: Ecology and Conservation*,
DOI 10.1007/978-1-4419-8468-5_8, 

which is available, the picture which emerges is a mixed one, with a decline in populations recorded from some sites and increase in numbers for others. At Keoladeo Ghana National Park, Bharatpur, which has been closely monitored by bird lovers, conservationists, and environmental activists over the years, a definite decline in the populations of heronry birds has been observed. At the turn of the previous century, the number of breeding pairs of Painted Stork in the extensive park was estimated conservatively to be around 4,000 (Ali 1953; Sankhala 1990). If all the other species of colonial waterbirds nesting in the park were included then the total number of breeding pairs would be in excess of 10,000. However, in the last decade the breeding pairs of all manner of colonial waterbirds recorded from here is considerably less (Venkitachalam and Vijayan 2009). Recent observations at Keoladeo, in October 2010, gave an impression that the total number of nests of the Painted Stork was not more than a few hundreds (A.J. Urfi, unpublished observations).

Regarding Kokkare Bellur, Neginhal (1977) felt that the village's Painted Stork colony was in a phase of expansion at the time of his visit in the late 1970s. According to local estimates, more than 1,000 pairs of pelicans were nesting at Kokkare Bellur in the 1980s. But a decade later the number of breeding pairs had come down to 160 (Subramanya and Manu 1996). According to another contemporary report, Painted Storks at the village ranged between 850 and 900 and pelicans between 300 and 350 nesting pairs (Sridhar and Chakravarthy 1995). The decline was attributed to tree felling, hunting, decline in food supplies, and pesticide poisoning (Manu and Jolly 2000).

However, at the Piele gardens and nearby Ghogha area, in the city of Bhavnagar the trend seems to be the reverse. According to Parasharya and Naik (1990) the total number of Painted Stork breeding pairs recorded from Peile gardens were 96 and 70 in 1980 and 1981, respectively. According to another report, in 1979 only 42 and 23 nests of Painted Stork were recorded from Piele gardens and Ghogha area, respectively. Later, while the Ghogha colony disappeared, those in Piele gardens grew in size (Anonymous 2005). In 2003, about 482 Painted Stork nests were recorded from here, and in 2004 the count was 328 nests. These figures suggest that overall there has been an increase in the numbers of Painted Stork at Bhavnagar city gardens. While it is held that the local municipal authorities have worked hard to maintain the city gardens and also ensure that visitors do not disturb the birds, but what are we to make of these trends? Are more Painted Stork nesting in Bhavnagar because, overall, Painted Stork are doing well and are becoming abundant or is it because in the country, as their natural breeding habitats are being lost, they are being forced to flock to the safe confines of gardens in cities, such as the gardens of Bhavnagar, which afford some degree of protection and also a suitable nesting habitat. The case of the Delhi Zoo colony, which will be discussed shortly, is somewhat similar. There is an impression that in South India too the populations of Painted Stork have increased over the years. Also, several new colonies, including a few in urban areas, have also come up (Rahmani in press).

## 8.2 Threat Factors

From a conservation viewpoint, heronry birds are more vulnerable, not just due to their specialized diet and foraging but also because of the concentration of their nesting in space and time. Many species on the endangered list include colonially nesting waterbirds (Birdlife International 2001). In the later sections, some of the threat factors with respect to heronry birds, with an emphasis on Painted Stork are discussed.

### *8.2.1 Habitat Loss*

Loss of feeding habitats or alterations in them are important factors behind the decline in wetland birds. All across Asia, wetlands, the principal foraging ground for birds like Painted Stork are being destroyed at an alarming rate. Unfortunately for them, a view prevails among administrators and land developers that wetlands are wastelands, breeding grounds for malaria and mosquitoes, and therefore fit only for land reclamation and developmental projects. With the wave of economic liberalization having swept many parts of Asia, India particularly, land has become a precious commodity as shopping malls, roads, highways, commercial complexes, etc. have mushroomed. Satellite townships around major metropolises have sprung up in the past decade due to which areas of wilderness, especially wetlands, have been a first casualty (Urfi 2010a).

While being a matter of concern for all manner of birds, including migratory waterfowl, habitat loss in the vicinity of heronries is particularly alarming because the feeding areas for birds disappear. A case in point is Sultanpur Bird Sanctuary. Here, as at other sites too, the birds nesting in this park freely move in and out of the park premises for foraging. From observations of their foraging flights, it appears that their foraging grounds lie towards the north of the park, at marshes associated with the Najafgadh drain, which in recent years have become the center of several land development projects. Towards the easterly direction lie the wetlands of Basai (Islam and Rahmani 2004). Recent reports (Narain 2009) suggest that much of the land in the neighborhood of Basai has already been earmarked for various development projects. With the construction of a flyway, the once small, sleepy village of Basai has started looking like a satellite of the city of Gurgaon. Meanwhile, Gurgaon itself has grown rapidly in size and because of its proximity to Delhi, become a hub of commercial activity.

Foraging areas for waders like the Painted Stork are also lost if natural wetlands are converted into fish farms. In many parts of India, natural wetlands, of the seepage type or rain water collection type, have been deepened and used for culture fishery. While this may be a boon for some type of piscivorous birds such as cormorants and kingfishers, they are quite useless for waders. Habitat alteration is a serious threat for the conservation of storks in other parts of the world too. For instance, in Australia due to the modification of habitat for the cultivation of cash crops, populations of Black-necked Stork have declined in many parts (Dorfman et al. 2001).

Catchment fishery at inland wetlands is also a serious threat for fish-eating birds because it depletes the fish stock leaving nothing for the birds. We have some idea about the quantities of fish harvested by fish-eating birds such as stork, pelican, cormorants, etc. during the period of their nesting (Chap. 6). At Kolleru Lake the decline in numbers of pelican was attributed to fishing in the local waters, among other factors (Guttikar 1978). The growing competition from humans for fish, leading to a severe reduction in fish supplies could probably be a major reason for the decline in numbers of these birds and abandonment of colonies at several sites across India.

Another cause for concern is the spread of exotic species of fishes. *Tilapia* for instance, was introduced in India during the 1950s (Chap. 6) and has spread all across the country and is regarded as a noxious invasive species. It was recorded in the diet of the Painted Stork recently (Kalam and Urfi 2008) but was not recorded in the 1960s (Desai et al. 1974). Recently, some reports in the Indian media drew attention to the spread of other species of exotic fishes in Indian fresh waters particularly at Sultanpur National Park. The impact of these invasive species on the ecology of wetlands needs to be investigated in a scientific manner.

### 8.2.2 *Pesticides and Toxins*

Although the problem of pesticides with respect to birds in general and fish-eating birds in Asia in particular has been extensively reviewed (e.g., Tanabe and Subramanian 2003), very few studies on this aspect have actually been done in India. However, colonially nesting fish-eating birds can serve as excellent bio-indicators and their study can be extremely rewarding from an environmental toxicology point of view. Their several advantages include the following. Primarily, their eggs can be examined easily which are also easy to collect and can be transferred to the lab for analysis. Secondly, the high lipid content of the avian egg (7–10%) makes it a good medium in which to measure lipid soluble environmental contaminants (Fox and Weseloh 1987).

Contamination of the soil and water bodies by pesticides and heavy metals is widespread in India. As several studies have pointed out, the degree of contamination of water bodies all across India, where pesticides used in agriculture reach by surface run off and leaching, is extremely high. The concentration levels and distribution pattern of the persistent organochlorine pesticide (OCPs) residues, including aldrin, dieldrin, endrin, HCB, HCH isomers, DDT isomers/metabolites, endosulfan isomers ($\alpha$ and $\beta$), endosulfan sulfate, heptachlor and its metabolites, $\alpha$-chlordane, $\gamma$-chlordane, and methoxychlor is, in many cases, way above the threshold level (Malik et al. 2009). High levels of organochlorine and organophosphate insecticide residues have been reported in tissues of fish and other animals (Singh et al. 2008). The Himalayas, it would be expected, would be relatively free of pesticide residues but here too organochlorine residues have been detected both in the water as well as tissues of aquatic organisms (Sarkar et al. 2003). It is notable that in spite of a complete ban on Aldrin and restrictions imposed on the usage of DDT and HCH in India (UNEP 2003) these pesticides continue to be used.

Pesticides act in insidious ways. If broadcast in high doses they directly affect the birds and the unintended target organism, causing abnormalities in behavior and even mortalities. For instance, deaths of birds, including those of the Sarus Crane, were recorded on the outskirts of Keoladeo Ghana, Bharatpur. (Banned pesticides continue to be used by farmers in the agricultural fields surrounding the park.) On examination of the carcasses it was discovered that the cause of death could be attributed to ingesting OCPs, along with the food outside the park (Muralidharan 2000).

Most pesticides are neurotoxic i.e., they effectively block the transmission of nerve impulses across neural synapses. Therefore, one of the visible affects of pesticide poisoning is changes in the motor behavior. Some years ago at Keoladeo I saw an adult Painted Stork behaving abnormally. Standing on a dyke along with some other birds, at the approach of some tourists while all the birds immediately took to the wing this bird was unable to take off. Finally, when it did take off it fluttered around, lost balance, and almost crashed into a tree (A.J. Urfi, unpublished observation). Certainly, this was abnormal behavior and something seemed to be wrong with its motor coordination. Since there were no visible signs of injury, possibly its erratic behavior could be attributed to pesticide poisoning.

Another way in which pesticides act is when they travel along the food chain, and at each trophic level gets progressively magnified. Initially in sublethal doses, by the time they have entered the body of a top level carnivorous predator along with the food, their concentrations are extremely high. Once inside the body of the bird they start interfering with its metabolism. One of the well known effects is when pesticides start affecting the egg shell metabolism by blocking the pathways leading to the metabolism of calcium. As is well known, this leads to production of eggs with thin shells which are liable to break before the embryo has developed fully. The results are population declines and mortalities of undeveloped embryos.

Since virtually no work has been done on colonial waterbirds in India, for the present we can only get an idea about the extent of the pesticide contamination problem by taking note of some other studies on birds. For instance, a study by BNHS during 1991–1996 revealed that in Corbett National park, in the Himalayan foothills, the Lesser-Fish Eagle were unable to breed successfully mainly due to egg shell thinning (Anonymous 1997).

### *8.2.3 Poaching and Egg Stealing*

Poaching for meat and eggs occurs at heronries all across India and is reckoned to be an important factor responsible for population declines in Painted Stork (Birdlife International 2001). According to several sources, young and eggs are taken from nests both for food as well as profit (Hancock et al. 1992). A good price is obtained by hunters, often professionals, for young birds to be kept as pets or sent to zoos. A few site-specific cases of poaching in India have been mentioned already in Chap. 7. In Pakistan, the population of Painted Stork is reported to be rapidly dwindling due to continuous poaching by fishermen who capture and sell its chicks to animal exporters (Hasan 2001).

### *8.2.4 Urbanization: A Case Study of Delhi*

Given that, a large number of Painted Stork colonies in India are located in urban premises (Subramanya 1996) a discussion about how urbanization impacts biodiversity, particularly heronry birds is of relevance. Urbanization is a frequently cited cause of biodiversity loss (Czech and Krausman 1997) but our understanding about urban ecology is severely limited (Marzluff et al. 2001). This is probably because several other aspects of urbanization such as health, housing, transport issues, etc., all of immense importance, tend to grab the immediate attention of authorities and researchers with the result that biodiversity issues often take a back seat. Also, while there seems to be general awareness that increasing urbanization is leading to biodiversity loss, much of the conservation effort is understandably focused on the wilderness areas.

An important aspect of urbanization is that due to the increasing spread of built up areas, natural areas, especially around the periphery of cities, are being lost. On the other hand, some types of birds, especially those species which exhibit a high degree of behavioral plasticity, are increasingly becoming dependent upon resources available in urban premises, such as nesting sites and food. Most probably this is happening because their natural habitats are being destroyed at a fast pace forcing them to flock to safe sites in urban parks and zoos premises.

Given the importance of the Delhi Zoo's Painted Stork population, and its dependence on surrounding areas of the river Yamuna floodplains, particularly the Okhla Bird Sanctuary, and also given that this system has been closely monitored, there is a need to discuss it in detail. Located about 9 km away from the Delhi Zoo, the Okhla Bird Sanctuary is an important feeding ground for the Painted Stork, but as Fig. 8.1 shows the area is increasingly becoming isolated due to increase in built up areas from all sides. A number of encroachments on the flood plain of the river Yamuna have happened over the past decade, often in disregard to environmental considerations. Consequently, changes with respect to bird life are already visible at Okhla. For instance, indicator species such as the Sarus Crane have not been sighted at Okhla since 1992 (Urfi 2006). Also, there has been a marked reduction in the diversity of waterfowl over the period of a decade (Fig. 8.2). Indeed, a decline in all diversity and abundance parameters of waterfowl at Okhla has been recorded (Table 8.1).

It is of course not possible to say conclusively that the increase in urbanization is directly responsible for the decline in bird diversity since several other factors may also be responsible. However, there seems to be a correlation between the increasing proportions of built up areas due to the construction of new bridges, flyways, housing

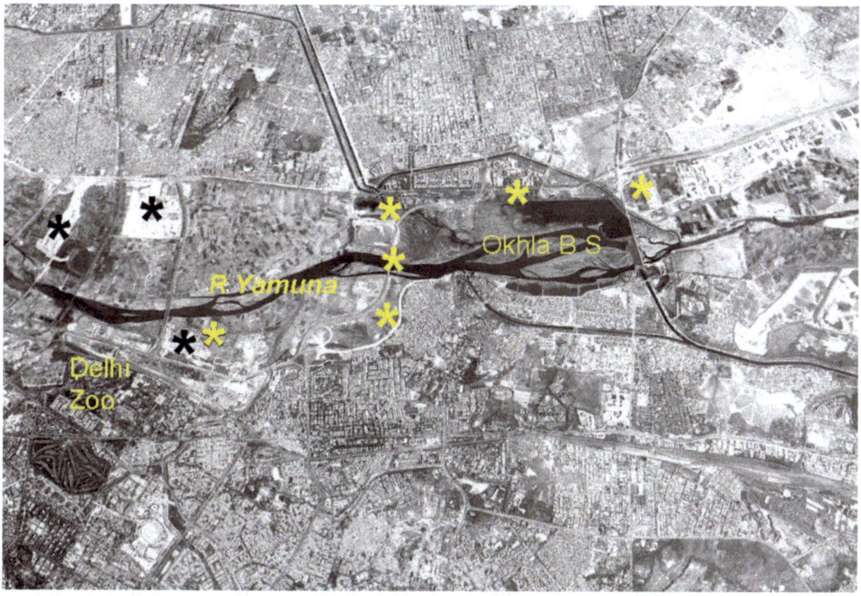


**Fig. 8.1** Satellite imagery of the river Yamuna floodplain in Delhi to show the location of Delhi Zoo in relation to river Yamuna and the Okhla Barrage Bird Sanctuary, the main feeding ground of Painted Stork. Note extensive encroachments on the river Yamuna flood plain and built up areas all around the river. *Asterisks* marks out built up areas which have come up in the past few years. *R Yamuna* River Yamuna; *Okhla B S* Okhla Bird sanctuary. *Source*: Google Earth (2010)

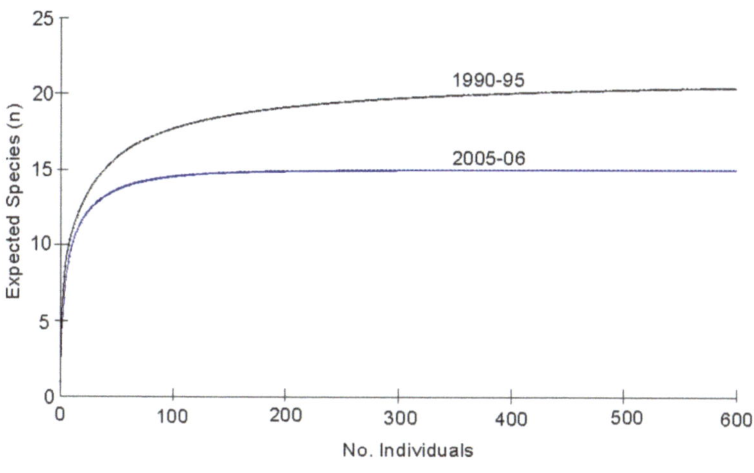


**Fig. 8.2** Cumulative species-individual relationships for waterfowl at Okhla Bird Sanctuary censured during 1990–1995 (*upper curve*) and 2005–2006 (*lower curve*). *Source*: Urfi (2006)

**Table. 8.1** A comparison of various ecological diversity parameters at Okhla Bird sanctuary during two distinct periods separated by almost one decade

| | Periods | |
|---|---|---|
| Parameters | 1990–1995 | 2005–2006 |
| Waterfowl abundance | 25,112 | 8,940 |
| No. species in sample | 27 | 20 |
| Diversity (*D*) | 5.711 | 3.022 |
| Shanon diversity | 2.008 | 1.581 |

*Source*: Urfi (2006)

projects, malls, super markets, etc., on the one hand and the decline in bird populations (Urfi 2010b). Additionally, as mentioned in several reports, the waters of the river Yamuna, where the Painted Stork catch their fish, are increasingly getting polluted. Traces of pesticides and heavy metals having been detected from the river (Rawat et al. 2003). Several drains from the city discharge their effluents into the river due to which approximately 20 km Delhi segment of the river Yamuna is the most polluted along the entire length of the river.

Given the deterioration of the ambient environment and the enormous pressure on bird habitats, the question is what of the Delhi Zoo and its wild waterbirds? Here too, the general impression is that the wild birds of the zoo campus have declined over the years (Malhotra 2006) but as far as the Painted Storks are concerned it is difficult to say with certainty if their populations have been adversely impacted. In the 1960s, when the population got established, the few published records available suggest that the populations then hovered around 400. The highest number recorded was 450 in 1967. Interestingly, during the period 1988–1992, the maximum number of birds recorded was 550 (in 1988–1989) (Urfi 1997). In recent years, count figures have been similar (Meganathan and Urfi 2009) which leads us to the conclusion that there has been no major change in the zoo Painted Stork population. Having said this, the picture looks slightly different if anecdotal records of the same period, which suggest that the number of Painted Stork then could well be in excess of 800 (A.K. Malhotra, personal communication) are taken into consideration. But assuming that there is an upward trend at the Delhi Zoo, as seems to be the case with the Bhavnagar population, how should this trend be interpreted? Does it mean that "all is well" in spite of the sweeping changes in the surrounding environment? Alternatively, since suitable nesting areas are getting scarce in the country, the birds have no option but to breed at sites like the zoo where they can find safety and easy access to foraging grounds. A conservation monitoring program to monitor select heronries, with the Delhi Zoo being a pilot study, needs to be initiated as early as possible (Urfi 2010b).

Clutch size is an important though controversial fitness parameter widely used in conservation biology. An interesting study, of great relevance from the point of view of Painted Stork, is the analysis of clutch size data of Wood Stork in Florida during the period 1875–1967 (Rodgers 1990). A careful analysis of clutches in museum collections revealed that although in the initial period the clutches were significantly larger compared to the 1960 and later, no overall declining trend was discernible.

As far as the Delhi Zoo Painted Stork population is concerned, the mean clutch size for each of the years from 1966 to 1971 ranged from 2.2 to 2.8 (mean = 2.49) (Desai et al. 1977). Interestingly, our studies during 2005 and 2006 (Meganathan and Urfi 2009) recorded a slightly lower range, 2.09–2.43. For Pond 1 (which were also studied by Desai et al. during the 1960s) the mean was 1.99 and range was 1.75–2.42, suggesting a decline in clutch size but only in Pond 1. However, since these observations are inconclusive and preliminary more studies are required.

### 8.2.5 Disturbance

In spite of the fact that several species of stork, ibis, spoonbill, egrets, herons, and cormorants build colonies in the midst of busy cities, giving the impression that these birds are tolerant of human presence, numerous other instances indicate that actually heronry birds are extremely sensitive to disturbance. The disturbed birds may either abandon nesting altogether or be forced to nest in places or situations which are sub-optimal, eventually compromising on fitness. At the Delhi Zoo there is a recorded instance when Painted Stork abandoned nesting altogether due to a suspected human disturbance factor. This happened in 1988 when at the commencement of the breeding season a fairly large number of birds settled down to breed. Coincidentally, during the same period the city of Delhi witnessed a massive political rally in which several thousand people came from the neighboring state of Haryana. According to reports, on the mentioned day, a large number of the rallyists poured into the zoo and created noise and disturbance (Urfi 1990a). The visitors were reported to have behaved in a disorderly manner, which the zoo authorities were unable to check. Just a day after the disturbance the Painted Stork, which had by this stage laid eggs, abandoned breeding and flew away not to return back that year (Urfi 1990a, 1997).

Other instances of disturbances influencing the nesting of storks are also on record. For instance, at Raiganj Sanctuary in the state of West Bengal, a study reported that the pressure of human disturbance close to the nesting trees, forced Asian Openbill to abandon their preferred trees and change to other trees and nest at a different height (Datta and Pal 1993).

Several other cases on how disturbance and human activities effect Painted Stork have recently been reviewed (Urfi 2010c). A brief mention of some interesting cases is given as follows. In Udupuria colony in Rajasthan, nestlings and juveniles were attacked by (honey) bees when a hive on one of the nesting trees was accidentally disturbed by villagers. Twelve nestlings and 23 subadults were found dead, upto 200 m from the colony (Nair 2006). At Bhavnagar, many subadults making their initial flights reportedly got entangled in kite strings and were killed. Unfortunately, the timing of kite flying festival in the city coincides with the time when the young are big enough to make their first local flights (Parasharya and Naik 1990). In addition to these, disturbances leading to nest abandonment, either due to the presence of large number of people near the colony or bursting firecrackers and putting up scare crows are also on record (Gadhvi 2002; Vashishtha 2001). At Ranganthittu when the tourist boats go very near the breeding birds, they get frightened and fly away leaving their nests unprotected. There is an interesting report of crows anticipating this situation, following the boats and pillaging the eggs and even taking away the nestlings from the unguarded nests (Neginhal 1982).

If disturbance manifests itself while the bird is foraging then foraging time is lost for which the bird may have to compensate by feeding for longer time and thereby compromising on other essential life activities. In the Delhi region we found that out of 127 foraging bouts recorded, 6% were those that were abruptly terminated by disturbance due to a human or cattle (Kalam and Urfi 2008). Interestingly, all disturbed bouts were recorded outside the protected areas indicating that disturbance in the open country can be a serious problem and needs to be factored into conservation policies.

## 8.3 People and Conservation

The usual approach in state sponsored conservation activities is that governmental agencies undertake land acquisition in order to set them aside as protected areas. The areas which they select are already areas of high biodiversity or ornithological significance, as in the case of "Important Bird Areas" (IBAs) (Islam and Rahmani 2004). The philosophy underlying this relies on the ecological approach, based on the theory of island biogeography and species-area relationships, which are the pillars of modern conservation ecology. Though there are controversies, of a scientific nature, from time to time, such as the one about several small protected areas vs. one large reserve, on the whole this approach is the dominant paradigm of conservation all over the world.

However, nothing is more serious than the problems linked with land acquisition and the resettlement of local communities who have been living in the area for centuries. Alternate thinking seeks to ask, why set aside areas for biodiversity by depriving indigenous people access to bio-resources that they have been using for generations? Why not allow for a more wholesome protection of the countryside? If the protected areas approach has to be adopted then why not invite the participation of local people, wholeheartedly? Although a full discussion about the social dimension of conservation is beyond the scope of this book, a few examples of relevance for wetlands and heronries are discussed here.

The controversy over cattle grazing that erupted at Keoladeo Ghana in 1980 is illustrative of the pitfalls of the traditional approach of managing protected areas. The importance of this park as a place of high biodiversity and conservation significance has already been discussed (Chap. 2) but reverse the coin and a totally different picture emerges because the park also happens to be the only grazing ground for thousands of cattle belonging to farmers from the nearby 14 villages. In the arid, hostile wilderness of the desert state of Rajasthan, competition for scarce resources, thus makes Keoladeo a battle ground for conservationists and villagers and is symbolic of the unique type of problems in conservation, faced in developing countries.

In the 1980s some well meaning scientists, ornithologists, and conservationists, headed by a team from BNHS under the stewardship of Dr. Salim Ali, appealed to the Government to enforce a ban on cattle grazing in an effort to make Keoladeo an inviolate space for birds, heronries and at that time, for the Siberian Crane. For centuries the villagers had been grazing their cattle in the park and there was no objection from the local administration but now with this ban, enforced with all power at the disposal of the state, the villagers were incensed. What happened subsequently is now all history (Urfi 1990b). There were numerous clashes between local people resisting the ban and the police and park officials in which several people died. A couple of years later several changes started taking place at Keoladeo itself. The scientists and conservationists in their enthusiasm for creating an "inviolate" space for the birds had overlooked the fact that in context of a wetland the activities of herbivores are highly significant. Indeed, to use ecological terminology, since the wetland is a transitional ecosystem, for it to remain so and not transform into a climax forest through ecological succession, some of man's activates such as cattle grazing are necessary. Indeed, over thousands of years, man and the ecosystem have coexisted, with man reaping the benefits of wetlands in a sustained manner. But what really happened after the ban on grazing was enforced was that in the absence of the grazing pressure a grass known as *Paspalum distichtum* grew everywhere, chocking the wetland and making it useless as a habitat for birds. Indeed, the BNHS hydrobiological study also reported that areas which had tall grass were the least suitable for Painted Stork nesting areas (Ali and Vijayan 1983) besides the fact that the grass infested wetlands now had a considerably reduced depth, which adversely impacted the populations of invertebrates and fishes. When cattle graze they trample over the soil and churn it, thereby enhancing the fertility of the soil. Now that benefit was also gone. Therefore, some years later, by the time the authorities realized their mistake and decided to lift the ban on cattle grazing, the famous wetland had become an unsuitable habitat for all manner of birds.

The Keoladeo story is illustrative of the pitfalls of "management" of water bird sanctuaries by not only scientists but also the custodians of law and order, who treat local people living in the immediate vicinity of protected areas, with disdain and of no significance. Kokkare Bellur also illustrates this dimension of the problem. After the "discovery" of the village for heronry birds in 1976, the state forest department took interest in preserving the site so that the nesting of storks and pelicans could continue uninterruptedly. In the beginning things went off well but soon some events took place in which the balance between the villagers and the state agencies began to get disturbed.

In 1982 when a national park was to be set up near Bengaluru, the forest department was in need of chicks of pelicans for release in the new reserve. They decided to obtain these from Kokkare Bellur but the way they went about it was in a rash manner. Instead of obtaining permission from the village Panchayat,[1] the forest officials, acting on their own, simply stole the chicks from the nests. This was seen as a breach of trust by the villagers and created bad blood (Manu and Jolly 2000).

[1] A village level body for taking decisions and settling disputes.

**Table 8.2** Karnataka state forest department compensation scheme for trees on which Painted Stork and pelican build nests at Kokkare Bellur

| | | Compensation offered | | Actual value | |
|---|---|---|---|---|---|
| Tree species | Utility | In Indian rupees | In US$[a] (approx.) | In Indian rupees | In US$[a] (approx.) |
| *Tamarindus indica* | Fruits | 600–700 | 13.3–15.5 | 2,000–3,000 | 44.4–66.6 |
| *Ficus benghalensis* | Fodder | 80–300 | 1.7–6.6 | 500–800 | 11.1–17.7 |
| *Ficus religiosa* | Fodder | 80–300 | 1.7–6.6 | 500–800 | 11.1–17.7 |
| *Thespesia populnea* | Fodder | 60–100 | 1.3–2.2 | 200 | 4.4 |
| *Accacia nilotica* | Fodder | 100 | 2.2 | 100 | 2.2 |
| *Delonix alata* | Fodder | 60 | 1.3 | 100 | 2.2 |
| *Azadirachta indica* | Fodder | 60 | 1.3 | 100 | 2.2 |

*Source*: Manu and Jolly (2000)

[a]1 US$ = (approximately) 45 Indian rupees according to the exchange rate prevailing in 2000 (Wikipedia 2010)

Sometime during the same decade the forest department also issued an order that the trees used by the birds for nesting be protected from being felled, unless they were diseased or dead. However, there were many violations of this order since many locally powerful people went unpunished (Manu and Jolly 2000). As a compromise, the forest department offered the villagers an annual compensation if they did not chop off trees. However, as Table 8.2 shows the compensation offered was extremely meagre. Also, according to the villagers view, the compensation scheme was unfairly administered since certain applicants were favored over others.

## 8.4 Captive Breeding and Rehabilitation of Nestlings

The Painted Stork has bred successfully in a number of zoos in India and abroad (e.g., Sakkarbaug Zoo in Gujarat, Devkar et al. 2006). Advances in captive breeding and management of storks have also been reviewed and guidelines about housing, diet, veterinary care, etc. are available (Coulter et al. 1989). Different species of *Mycteria* are not difficult to breed in captivity provided some of their basic requirements have been met with. Since all of them are colonial nesters and usually nest high in trees surrounded by water it is therefore recommended that the enclosures for full-flighted birds be large enough for them to exercise their wings and stretch them full length. As far as the enclosure is concerned its height is more important than its length or width. Storks are very curious animals and may ingest sharp objects or particles that may cause impaction. Therefore, it is important that their enclosure be free of any such harmful objects. Special care has to be taken about the food provided to storks held under captive conditions. Providing unsupplemented meat or fish diets may cause calcium deficiency or a calcium/phosphorous imbalance. Vitamin E deficiency occurs in birds being fed primarily fish diets which are poor in quality, improperly stored, or high in unsaturated fatty acids.

**Fig. 8.3** Pelicans and Painted Stork taking advantage of the artificial (metal) structures provided by the forest department for nesting at Uppalapadu. Photo: Mrutyumjaya Rao

At many outdoor/open sites, arrangements for inducing wild populations of Painted Stork, pelican, and other species have been successful. Two cases which come to mind foremost are Negra zoo in Malaysia and Uppalapaddu in the South Indian state of Andhra Pradesh. In the case of Uppalapaddu, artificial breeding platforms made out of metal were installed in the tank which soon attracted pelicans and Painted stork (Fig. 8.3).

Rehabilitation of nestlings which fall off from nests has been undertaken by several local conservation groups and like-minded individuals from different parts of India. There is no doubt that such activities are extremely helpful, though their impacts need to evaluated. A few examples of such projects are given here. Bhadalwadi Tank has a large colony of Painted Stork and other species. Since the tank is under the control of the state irrigation department, the water is drawn from it as and when required for the purpose of irrigation. Some years ago, when the breeding of Painted Stork was in full swing, the water levels in the tank dropped suddenly with the result that the nests and their occupants immediately became easy prey for ground predators such as common mongoose, stray dogs, jackals, and even wolf. These predators which were previously kept in check by the water barrier around the nests could now be approached with ease. Several juvenile storks, partially eaten, were seen lying on the ground. A local environmental organization immediately got into action and attempted to save as many nestlings as possible, with the help of the state forest department and well known NGO's such as the BNHS (Pande 2006). About 200 nestlings were saved due to these efforts.

At Kokkare Bellur a similar project was undertaken to protect the nestlings that fall off from nests and are devoured by village dogs. Safety nets at some points to arrest the fall of nestlings if and when they fell off from nests were put up at several strategic locations by volunteers of a local NGO and the villagers (Subramanya and Manu 1996). Similar rehabilitation projects have been undertaken on other storks as well (Singha et al. 2006).

## 8.5 Conservation Solutions

Over the past few decades, much work and thinking has gone into thrashing out solutions for the conservation of storks worldwide (Luthin 1987). Many of the solutions are general i.e., about protection of habitats, checking poaching and egg stealing, etc. For these, there are laws already in place which need to be implemented properly. Simultaneously, there is a definite need for more research in ecology, distribution, and captive breeding of different species of storks. However, a few areas which merit a special discussion in context of conservation of heronry birds are discussed here.

### *8.5.1 "Liquid Gold"*

Much has been written about the fertilizing properties of waters surrounding heronries (Paulraj 1988). These waters, rich in nutrients, flow into the surrounding agricultural fields and enrich the soil. These heronry waters have been referred to as "liquid gold" and their use, in place of inorganic fertilizers, has been encouraged. In fact, some authors have gone on to suggest that their use as a panacea for solving the twin problems of increasing crop yields, propping up the local economy, and achieving the objectives of conservation, with the help of local village folk, as a bonus.

Paulraj and Kondas (1987) studied how the yields of crops varied in different types of treatments: with and without reservoir waters and with and without additional (chemical) fertilizers at Vedanthangal Bird Sanctuary. The results of their experiments are shown in Table 8.3. While an increase in yield after using reservoir waters compared to using only well water (devoid of any fertilizers) was noted, the increase seems marginal. However, if extra fertilizer is added then the difference in yield is much greater, from a mean of 3,575 kg/ha (dry-weight) using, reservoir – water without fertilizer to 5242.5 kg/ha (dry-weight), using reservoir – water with fertilizer. So, the results clearly show that the reservoir water has some benefit but probably not enough to make economic worth. Also the high yields that would bring the farmer some profit are possible only by the addition of extra fertilizer which, if it is a chemical one, then brings us back to square one as far as environmental protection is concerned.

**Table. 8.3** Yield of paddy grain at four experimental plots near Vedanthangal Bird sanctuary after treating them with well – water and reservoir water, with and without chemical fertilizers

| | Grain yield in kg/ha (dry-weight) from different treatments | | | |
|---|---|---|---|---|
| Subplot no. | Well water without fertilizer | Reservoir – water without fertilizer | Well water with fertilizer | Reservoir – water with fertilizer |
| 1 | 3,210 | 3,670 | 4,690 | 5,050 |
| 2 | 3,260 | 2,500 | 4,850 | 5,330 |
| 3 | 3,290 | 3,510 | 4,750 | 5,330 |
| 4 | 3,420 | 3,620 | 4,940 | 5,260 |
| Mean ± (sd) | 3,295 (77.6) | 3,575 (72.3) | 4807.5 (95.5) | 5242.5 (114.8) |

*Source*: Paulraj and Kondas (1987)

## *8.5.2 Environmental Education*

Heronries are an excellent resource material for nature education. They offer interesting opportunities for studying and appreciating the ecology of birds at one of the most significant phases in their life cycle i.e., the breeding phase (Urfi and Nareshwar 1998). Organizing heronry appreciation programs by volunteer bird-watchers and nature lovers can go a long way in sensitizing people about the biological, conservational and ecological aspects of heronries, and about biodiversity issues in general.

In Chap. 1 I began by asking the question why are the Painted Stork so called? Although I have no answer to this question, during the course of field work all across India I discovered that among all the Indian storks, this species is perhaps the most drawn and painted. In the vicinity of important heronries, such as Kokkare Bellur, where grass-roots level organizations are running environmental education programs in schools, one finds drawings of Painted Stork made by school children and their teachers (Fig. 8.4). Thousands of kilometers away, at Udpuria in Rajasthan state, school children in a painting competition held during the Wildlife Week, decided to make a colorful picture of a Painted Stork (Vyas 2009). Therefore, the Painted Stork itself is a valuable tool for nature education among school children especially from rural areas across India.

One of the most pressing needs is to develop environmental education and interpretation facilities. Opportunities for setting up educational programs at urban heronries may be the easiest but, since urban citizens also have a life style which involves a much greater use (and wastage) of energy and other resources, and given the fact that they take much of their privileges for granted, the need for them to be sensitized to nature conservation issues is paramount. However, as things stand, utilizing urban heronries as a resource for conservation education seems to be an opportunity lost in most cases, in India.

Generally speaking, environmental education does not figure very high on the agenda of most Indian zoos and the wild waterbirds on their campuses less so (Urfi 1996). At the Delhi Zoo, while some limited information is available about the wild

**Fig. 8.4** A painting of Painted Stork made by school teachers and students in a building near Kokkare Bellur village. Photo: A.J. Urfi (2001)

waterbirds nesting in the zoo premises, overall the wayside and other information facilities leave much to be desired (Fig. 8.5). The Delhi Zoo Painted Stork colony though it has been the subject of several studies, including those by one of its ex-directors, has not been highlighted by the authorities concerned. There is no pamphlet, educational material about it; the story of the Painted Stork, how they started arriving here in the 1960s; the fact that they come here year after year and enrich our lives, etc. needs to be told, and told properly. The Painted Stork nesting in the Delhi Zoo is a living resource and several interesting aspects of their biology can be a demonstration of basic biology studies for school and college children, who should also be involved in the study and conservation of these birds. Till date the interpretative waysides besides the zoo ponds where the Painted Stork nesting colonies are located are fairly basic in nature and in many cases in dire need of repairs and renovation (Fig. 8.5b). One wonders how far the objectives of sound nature education can be met with such simple waysides which merely aim to acquaint the observer with the bird (Fig. 8.6). Besides sprucing up their educational facilities, there is an urgent need for the concerned authorities to tap the energy and involvement of scores of bird-watchers and nature lovers in schools and colleges, by initiating heronry monitoring programs.

**Fig. 8.5** The condition of existing interpretative waysides near the Painted Stork colonies at the Delhi Zoo. (**a**) Shows Pond 1 colony with a panel of waysides of different species of wetland birds. (**b**) Shows the stump of a wayside near Pond 2. Photo: A.J. Urfi (2010)

### *8.5.3 Heronry Monitoring*

Birds are popular bio-indicators of environmental change because they are ecologically versatile, their populations as well as select fitness parameters can be conveniently monitored, often with the voluntary involvement of local nature enthusiasts across large geographical scales, and their presence/absence in a particular area is

**Fig. 8.6** Details of the wayside on Painted Stork in front of Pond 1 colony of the Delhi Zoo. Note the visual elements constitute a distribution map (*top right*) and a color drawing of the Painted Stork (*bottom, right*). The interpretation text is in Hindi (*upper*) and English. Photo: A.J. Urfi (2010)

consequential (Bibby et al. 1992; Urfi 2004). Therefore, the opportunities for meaningful ecological research offered by the wild waterbirds nesting in urban premises are immense and long-term conservation monitoring programs, involving volunteers and bird-watchers, to create data bases that will be useful for understanding urbanization, and climate change impacts on biodiversity that can be easily undertaken (Urfi 2010b).

Restricting our focus to the Delhi Zoo which has been well studied it is recommended that a program to monitor this population should be initiated and built into the operational machinery of the zoo. While being coordinated by the zoo authorities, under the guidance of researchers and scientists from academic institutions, this program could derive its inputs from the efforts of volunteers such as local bird-watchers and school and college students. This will set the tone for generating long-term data bases for breeding birds and also simultaneously taking a lead in initiating citizen science programs for the benefit of environment (Urfi 2004; Urfi et al. 2005; Greenwood 2007). Since the birds in question are fairly large sized and their nests easily approachable they are also easy to monitor and no special equipment is required to study them.

Long-term heronry monitoring has been extremely valuable in generating data for population studies. A case in point is the Heronries Census of the British Trust of Ornithology which began in 1928 and is today the longest-running breeding season monitoring scheme in the world (British Trust for Ornithology (BTO) 2009). If a similar program is initiated at the Delhi Zoo then its scientific advantages would be several. It will be useful in understanding population ecology issues, i.e., how various biotic and abiotic factors influence bird populations. In view of the recent concern about global climate change affecting the monsoon, it will help us understand better how yearly rainfall patterns effect Painted Stork nesting since these birds are monsoon dependant for breeding. Above all, this monitoring program would also help us to keep tabs on urbanization and pollution and their possible impacts on birds.

## References


Ali S (1953) The Keoladeo Ghana of Bharatpur (Rajasthan). J Bombay Nat Hist Soc 51:531–536

Ali S, Vijayan VS (1983) Hydrobiological (ecological) research at Keoladeo National Park, Bharatpur (first interim report). Bombay Natural History Society, Mumbai

Anonymous (1997) DDT threatens Lesser Fish-Eagle in Corbett. Hornbill 3:32

Anonymous (2005) Editorial. Flamingo 3:1

Bibby C, Burgess ND, Hill DA (1992) Bird census techniques. Academic, Oxford

BirdLife International (2001) Threatened birds of Asia: the birdlife international red data book. BirdLife International, Cambridge

British Trust for Ornithology (BTO) (2009) Heronries surveys. British Trust for Ornithology, Nunnery, Thetford. http://www.bto.org/survey/heron.htm. Accessed 10 Aug 2009

Coulter MC, Balzano S, Johnson RE, King CE, Shannon PW (1989) Conservation and captive management of stork. In: Proceedings of an international workshop at the New York's Zoological Society's Wildlife Survival Center, Saint Catherine's Island. Stork Interest Group, Georgia

Czech B, Krausman PR (1997) Distribution and causation of species endangerment in the United States. Science 277:1116–1117

Datta T, Pal BC (1993) The effect of human interference on the nesting of the Openbill Stork *Anastomus oscitans* at the Raiganj wildlife sanctuary, India. Biol Conserv 64:149–154

Desai JH, Shah RV, Menon GK (1974) Diet and food requirement of painted stork at the breeding colony in the Delhi zoological park. Pavo 12:13–23

Desai JH, Menon GK, Shah RV (1977) Studies on the reproductive pattern of the Painted Stork, *Ibis leucocephalus*, Pennant. Pavo 15:1–32

Devkar RV, Buch JN, Khanpara PS, Katara RD (2006) Management and monitoring of captive breeding of Painted Stork and Eurasian Spoonbill in zoo aviary. Zoos Print J 21:2189–2192

Dorfman EJ, Lamont A, Dickman CR (2001) Foraging behaviour and success of black-necked storks (*Ephippiorhynchus asiaticus*) in Australia: implications for management. Emu 101:145–149

Fox GA, Weseloh DV (1987) Colonial waterbirds as bio-indicators of environmental contamination in the Great Lakes. In: Diamond AW, Fillion FL (eds) The value of birds. Technical Publication No. 6. ICBP, Cambridge

Gadhvi I (2002) Painted Storks abandon nesting colony at Bhavnagar, Gujarat. Newslett Birdwatchers 42:12

Google Earth (2010) http://www.google.com/earth/index.html. Accessed 13 Nov 2010

Greenwood JJD (2007) Citizens, science and bird conservation. J Ornithol 148(Suppl 1):77–124

Guttikar SN (1978) Lost pelicanry. J Bombay Nat Hist Soc 75:482–485

Hancock JA, Kushlan JA, Kahl MP (1992) Storks, ibises and spoonbills of the world. Academic, London

Hasan M (2001) Birds of the Indus. Oxford University Press, Pakistan, Karachi

Islam MZ, Rahmani AR (2004) Important bird areas in India: priority sites for conservation. Bombay Natural History Society, Mumbai

Kalam A, Urfi AJ (2008) Foraging behaviour and prey size of the painted stork (*Mycteria leucocephala*). J Zool (Lond) 274:198–204

Luthin CS (1987) Status of and conservation priorities for the world's stork species. Colon Waterbirds 10:181–202

Malhotra AK (2006) Vanishing free living fauna of National Zoological Park, New Delhi. Indian Zoo Year Book 4:158–162

Malik A, Ojha P, Singh KP (2009) Levels and distribution of persistent organochlorine pesticide residues in water and sediments of Gomti River (India) – a tributary of the Ganges River. Environ Monit Assess 148:421–435

Manu K, Jolly S (2000) Pelicans and people: the two-tier village of Kokkare Bellur, Karnataka, India. Community based conservation in South Asia: case study no. 4. Kalpavriksh and International Institute of Environment and Development, New Delhi

Marzluff JM, Bowman R, Donnelly R (eds) (2001) Avian ecology and conservation in an urbanizing world. Kluwer Academic Publishers, Boston

Meganathan T, Urfi AJ (2009) Inter-colony variations in nesting ecology of Painted Stork (*Mycteria leucocephala*) in the Delhi Zoo (North India). Waterbirds 32:352–356

Muralidharan S (2000) Organochlorine residues in the waters of Keoladeo National Park, Bharatpur, Rajasthan. Bull Environ Contam Toxicol 65:35–41

Nair AK (2006) Udpuria – a stork paradise. Hornbill July–September:32–33

Narain V (2009) Growing city, shrinking hinterland: land acquisition, transition and conflict in peri-urban Gurgaon, India. Environ Urban 21:501–512

Neginhal SG (1977) Discovery of a pelicanry in Karnataka. J Bombay Nat Hist Soc 74:169–170

Neginhal SG (1982) The birds of Ranganathittu. J Bombay Nat Hist Soc 79:581–593

Pande S (2006) Bhadalwadi tank, a refuge for Painted Storks. Hornbill April–June:11–15

Parasharya BM, Naik RM (1990) Ciconiiform birds breeding in Bhavnagar city, Gujarat. In: Daniel JC, Serrao JC (eds) Conservation in developing countries: problems and prospects. Bombay Natural History Society, Mumbai

Paulraj S (1988) Impact of guano deposition in Vedanthangal water-bird sanctuary (Chengalpattu District, Tamil Nadu). J Bombay Nat Hist Soc 85:319–324

Paulraj S, Kondas S (1987) Two centuries of protection of a waterbird colony: how it may benefit the protectors. Environ Conserv 14:363–364

Rahmani AR (in press) Threatened birds of India and their conservation requirements. Bombay Natural History Society, Mumbai

Rawat M, Moturi MCZ, Subramaniam V (2003) Inventory compilation and distribution of heavy metals in wastewater from small-scale industrial areas of Delhi. Ind J Environ Monit 5:906–912

Rodgers JA Jr (1990) Breeding chronology and clutch information for the Wood Stork from museum collections. J Field Ornithol 61:47–53

Sankhala K (1990) Gardens of God. The waterbird sanctuary at Bharatpur. Vikas Publishing House, New Delhi

Sarkar UK, Basheer VS, Singh AK, Srivastava SM (2003) Organochlorine pesticide residues in water and fish samples: first report from rivers and streams of Kumaon Himalayan region, India. Bull Environ Contam Toxicol 70:485–493

Singh PB, Singh V, Nayak PK (2008) Pesticide residues and reproductive dysfunction in different vertebrates from north India. Food Chem Toxicol 46:2533–2539

Singha H, Goswami SK, Phukan R, Talukdar BK (2006) Rehabilitation of captive-reared greater adjutants *Leptoptilos dubius* in Assam. J Bombay Nat Hist Soc 103:315–320

Sridhar S, Chakravarthy AK (1995) Spotbilled pelicans and Painted Storks at Kokre Bellur – an update. Newslett Birdwatchers 35:22

Subramanya S (1996) Distribution, status and conservation of Indian heronries. J Bombay Nat Hist Soc 93:459–480

Subramanya S, Manu K (1996) Saving the spot-billed pelican. A successful experiment. Hornbill 2:2–6

Tanabe S, Subramanian A (2003) Bioindicators suitable for monitoring POPs in developing countries. In: Scientific and Technical Advisory Panel (STAP) and Global Environment Facility (GEF) workshop on the use of bioindicators, biomarkers and analytical methods for the analysis of POPs in developing countries, 10–12 Dec 2003. Ministry of Environment, Japan

UNEP (2003) Global report on regionally based assessment of persistent toxic substances. UNEP Chemicals, Geneva

Urfi AJ (1990a) Mysterious disappearance of Painted Stork from Delhi Zoo heronries and abrupt termination of their breeding. Newslett Birdwatchers 30:3–5

Urfi AJ (1990b) Combating the human factor. New Scientist, 26 May 1990, pp 68–69

Urfi AJ (1996) One bird's prison. New Scientist, 24 August 1996, p 45

Urfi AJ (1997) The significance of Delhi Zoo for wild waterbirds, with special reference to the Painted Stork *Mycteria leucocephala*. Forktail 12:87–97

Urfi AJ (2004) Birds: beyond watching. Universities Press, Hyderabad

Urfi AJ (2006) Biodiversity conservation in an urban landscape. A case study of some important bird areas on the river Yamuna in Delhi (India). In: McNeely JA, McCarthy TM, Smith A, Olsvig-Whittaker L, Wikramanayake ED (eds) Conservation biology in Asia. Society for Conservation Biology (Asia Section) and Resources Himalaya Foundation, Kathmandu, Nepal

Urfi AJ (2010a) Musings on a wetland. Jalaplavit 2:4–6

Urfi AJ (2010b) Using heronry birds to monitor urbanization impacts: a case study of Painted Stork *Mycteria leucocephala* nesting in the Delhi Zoo, India. Ambio 39:190–193

Urfi AJ (2010c) Factors causing nest losses in the Painted Stork (*Mycteria leucocephala*): a review of some Indian studies. J Bombay Nat Hist Soc

Urfi AJ, Nareshwar M (1998) Interpreting a village pond heronry. NewsEE (CEE) 4/3:5

Urfi AJ, Sen M, Kalam A, Meganathan T (2005) Counting birds in India: a review of methodologies and trends. Curr Sci 89:1997–2003

Vashishtha SC (2001) Painted Storks abandon colony after bursting of crackers at Pandad, Gujarat. Newslett Birdwatchers 41:13

Venkitachalam R, Vijayan L (2009) The current status of heronry population in Keoladeo National Park, Bharatpur. Newslett Birdwatchers 49:49–50

Vyas R (2009) Why Udpuria Painted Stork (*Mycteria leucocephala*) nesting colony should be declared a community reserve? Newslett Birdwatchers 49:43–46

Wikipedia (2010) Tables of historical exchange rates. http://en.wikipedia.org/wiki/Tables_of_historical_exchange_rates_to_the_USD#Table_for_recent_years. Accessed 2 Aug 2010

# Appendix

English and scientific names of animals and plants used in the book.

## Plants

Mesquite *Prosopis juliflora*
Neem *Azadirachta indica*
Peepul/The Sacred Fig *Ficus religiosa*
Tamarind *Tamarindus indica*
Indian Banyan *Ficus benghalensis*

Elephant Grass *Typha angustata*

## Vertebrates

### *Amphibians and Reptiles*

Mugger Crocodile *Crocodylus palustris*
Bengal monitor/Common Indian Monitor *Varanus bengalensis*
Malayan Water Monitor Lizard *Varanus salvator*
Checkered Keelback/Asiatic Water Snake *Xenochrophis piscator*

### *Birds*

Darter *Anhinga melanogaster*
Little Cormorant *Phalacrocorax niger*

A.J. Urfi, *The Painted Stork: Ecology and Conservation*,
DOI 10.1007/978-1-4419-8468-5, 

Indian Cormorant *Phalacrocorax fuscicollis*
Great Cormorant *Phalacrocorax carbo*
Little Egret *Egretta garzetta*
Western Reef Heron *Egretta gularis*
Great Egret *Casmerodius albus*
Intermediate Egret *Mesophoyx intermedia*
Cattle Egret *Bubulcus ibis*
Indian Pond Heron *Ardeola grayii*
Black-crowned Night Heron *Nycticorax nycticorax*
Snowy Egret *Egretta thula*
Grey Heron *Ardea cinerea*
Black-headed Ibis *Threskiornis melanocephalus*
White Ibis *Eudocimus albus*
Eurasian Spoonbill *Platalea leucorodia*
Dalmatian Pelican *Pelecanus crispus*
Spot-billed Pelican *Pelecanus philippensis*
Wood Stork *Mycteria americana*
Milky Stork *Mycteria cinerea*
Painted Stork *Mycteria leucocephala*
Yellow-billed Stork *Mycteria ibis*
Asian Openbill *Anastomus oscitans*
African Openbill *Anastomus lamelligerus*
Black Stork *Ciconia nigra*
Woolly-necked Stork *Ciconia episcopus*
European White Stork *Ciconia ciconia*
Abdim's Stork *Ciconia abdimii*
Storm's Stork *Ciconia stormi*
Maguari Stork *Ciconia maguari*
Oriental White Stork *Ciconia boyciana*
Saddlebill *Ephippiorhynchus senegalensis*
Black-necked Stork *Ephippiorhynchus asiaticus*
Jabiru *Jabiru mycteria*
Lesser Adjutant *Leptoptilos javanicus*
Greater Adjutant *Leptoptilos dubius*
Marabou *Leptoptilos crumeniferus*
The Shoebill *Balaeniceps rex*

Black Kite *Milvus migrans*
Spotted Eagle *Aquila clanga*
Steppe Eagle *Aquila nipalensis*
Imperial Eagle *Aquila heliaca*
Pallas's Fishing Eagle *Haliaeetus leucoryphus*
European White Stork *Ciconia ciconia*
Lesser-Fish Eagle *Icthyophaga humilis*
Peregrine Falcon *Falco peregrinus*

Shahin Falcon *Falco peregrinator*
Turkey Vulture *Cathartes aura*
Sarus Crane *Grus antigone*
Sandhill Crane *Grus canadensis*
Siberian Crane *Grus leucogeranus*
Lesser Flamingo *Phoeniconaias minor*
Painted Snipe *Rostratula benghalensis*
Common Curlew *Numenius arquata*
Bar-tailed Godwit *Limosa lapponiaca*
European Oystercatcher *Haematopus oestralegus*
Fish Crow *Corvus ossifragus*
House Crow *Corvus splendens*
Jungle Crow *Corvus macrohynchos*
European Starling *Sturnus vulgaris*

## *Mammals*

Raccoon *Procyon lotor*
Common Mongoose *Herpestes edwardsi*
Indian Jackal *Canis aureus indicus*
Gray wolf *Canis lupus*
Bonnet Macaque *Macaca radiata*
Rhesus Macaque *Macaca mulatta*
Blackbuck *Antilope cervicapra*

# About the Author

**Abdul Jamil Urfi** is an Associate Professor in the Department of Environmental Biology, University of Delhi, New Delhi, India. He studied fish physiology for his Ph.D., but subsequently switched over to field Ornithology. He did post-doctoral research on the behavioral ecology of the Oystercatcher at NERC ITE Furzebrook in England. Later, in India, he was involved with environmental education as a staff of CEE, Ahmedabad, where he also carried out a number of studies on reptiles and snakes. Painted Stork has been the focus of his research for the past two decades. Besides several research papers and articles, he has also written two books, *Birds Beyond Watching* (Universities Press, India) and *Birds of India, A Literary Anthology* (Oxford University Press, India).

# Index

MIX
Papier aus verantwortungsvollen Quellen
Paper from responsible sources
FSC® C105338

If you have any concerns about our products,
you can contact us on
**ProductSafety@springernature.com**

In case Publisher is established outside the EU,
the EU authorized representative is:
**Springer Nature Customer Service Center GmbH**
**Europaplatz 3, 69115 Heidelberg, Germany**

Printed by Libri Plureos GmbH
in Hamburg, Germany